Women and Men of the Engineering Path: A Model for Analyses of Undergraduate Careers

Clifford Adelman
Senior Research Analyst
Office of Educational Research and Improvement
U.S. Department of Education

For sale by the U.S. Government Printing Office
Superintendent of Documents, Mail Stop: SSOP, Washington, DC 20402-9328
ISBN 0-16-049551-2

U.S. Department of Education
Richard W. Riley
Secretary

Office of Educational Research and Improvement
Ricky T. Takai
Acting Assistant Secretary

**National Institute on Postsecondary Education,
   Libraries, and Lifelong Learning**
Carole Lacampagne
Director

Media and Information Services
Cynthia Hearn Dorfman
Director

National Science Foundation
Neal F. Lane
Director

National Institute for Science Education
Andrew Porter and Terrence Millar
Co-Directors

May 1998

The research reported in this monograph was supported, in part, by an interagency agreement between the National Science Foundation and the United States Department of Education (No. 9453866), and by a cooperative agreement between the National Science Foundation and the University of Wisconsin-Madison (Cooperative Agreement No. RED-9452971). At the University of Wisconsin-Madison, the National Institute for Science Education is housed in the Wisconsin Center for Education Research and is a collaborative effort of the College of Agricultural and Life Sciences, the School of Education, the College of Engineering, and the College of Letters and Science. The collaborative effort is also joined by the National Center for Improving Science Education, Washington, DC. Any opinions, findings, or conclusions are those of the author and do not necessarily reflect the views of the supporting agencies.

Contents

Acknowledgments

This document is the third to emerge from a cooperative relationship between the Office of Educational Research and Improvement (OERI) of the U.S. Department of Education and the National Science Foundation (NSF). *Women and Men of the Engineering Path* owes much to the support of NSF's Division of Science Resource Studies, and to the encouragement and feedback of the "College Level I" team of NSF's National Institute for Science Education (NISE) at the University of Wisconsin-Madison. I am particularly grateful to Mary Golladay of NSF and Prof. Arthur Ellis of the Department of Chemistry at UW-Madison for their faith in and patience with the production of this study. These relationships demonstrate how federal agencies and the research centers they sponsor can work together to address problems of common concern.

The idea of exploring the "engineering path" through an empirical examination of a national college transcript sample was presented first to the "College Level I" team of NISE in 1995. Sample transcripts, tentative decision-rules for classification, and a preliminary taxonomy were set before the team, and indispensable suggestions for the subsequent investigation were offered, particularly by Denise Denton, Dean of the School of Engineering at the University of Washington, and Sheilah Tobias.

The most critical preliminary task consisted of the line-by-line reading of college transcript records for 8,395 students in the High School & Beyond/Sophomore Cohort to determine engineering path status. The task took two readers, and without the assistance of Catherine Mobley (currently an Assistant Professor at Clemson University), the assignment of engineering path destinations would have been neither possible nor reliable.

In the course of many drafts of this study, I have received helpful feedback and suggestions from reviewers: Elaine Seymour of the University of Colorado, Jerilee Grandy of the Educational Testing Service, Mike Corradini of the College of Engineering at the University of Wisconsin-Madison, Dennis Carroll of the National Center for Education Statistics, and my colleagues Nevzer Stacey, Sandra Garcia, Carole Lacampagne, and especially Harold Himmelfarb. Thanks, too, to special assistance along the way from Maresi Nerad of the University of California/Berkeley, Charles Ratliff of the California Postsecondary Education Commission, and most of all to my younger son, Nicholas Adelman, who began his college career as a chemical engineering major and is finishing in chemistry, for truly enlightening discussions along the path he discovered.

Lastly, all of us working in the vineyards of higher education owe more than some will ever admit to the National Center for Education Statistics (NCES), which had the wisdom to include national college transcript samples in its longitudinal studies, thus providing a resource of knowledge that is simply unique in this world. With NCES came not only the data, but also my extraordinary colleagues and teachers, Nabeel Alsalam (now at the Congressional Budget Office) and C. Dennis Carroll. Without them, our national knowledge would be so much less.

Executive Summary

This monograph seeks to provide college academic administrators, institutional researchers, professional and learned societies, and academic advisers with a tapestry of information to improve their understanding of the paths students take through higher education. It begins with the observation that of those students who earn bachelor's degrees by age 30, 16 percent entered with no particular major in mind, and only 42 percent of the balance earned degrees in their intended field. These data indicate a considerable degree of student field migration.

The study demonstrates that migration rates are by-products of factors in students' choice of field, including curricular momentum and quality of academic performance carried forward from high school, the growing trend toward multi-institutional attendance, the nature of community college curricula for transfer students, credit loads and stop-out behavior, classroom experiences, changing student perceptions of the labor market, and student misconceptions of what given fields of study and occupations are all about.

Engineering was chosen as a case because it brings all the variables affecting choice, persistence, and migration into play. And because undergraduate engineering programs are offered in a limited number of institutions, we can offer a sharper primary story line about student history and choice. Engineering was also chosen because, while the overall "attrition" from the field is not high after students reach the "threshold" of the field, it is much higher for women than men, an unfortunate situation in a discipline with a historically severe gender imbalance.

The evidence used in *Women and Men of the Engineering Path* comes principally from the 11-year college transcript history (1982–1993) of the High School & Beyond/Sophomore Cohort Longitudinal Study (HS&B/So), as well as the high school transcripts, test scores, and surveys of this nationally representative sample.

This is the first national tracking study of students in any undergraduate discipline that identifies attempted major fields from the empirical evidence of college transcripts. A "curricular threshold" of engineering was defined, and the careers of students described with reference to that threshold. While 16 long-term "destinations" of students who reached the threshold are identified, they are collapsed into four for purposes of analysis:

- Thresholders, who never moved beyond the requisite entry courses.

- Migrants, who crossed the threshold of the engineering path, began to major in engineering, but switched to other fields or left college altogether.

- Completers, some of whom continued on to graduate school by age 30.

- Two-year-only students, whose college experience was confined principally to engineering tech programs in community colleges.

Selected Findings

Attendance Patterns and Degree Completion.

● Attending more than one institution is not a drag on degree-completion—for anyone. More than half of the HS&B/So college students attended more than one college, and 40 percent of this group crossed state lines in the process.

● Community college transfer students evidence strong preparation, with degree completion rates equivalent to those of 4-year college students. The transfers constitute 1/6th of the degrees awarded in engineering.

● The bachelor's degree completion rates (in any field) of students who reach at least the threshold of the engineering path are much higher than those for anybody else.

● While there is a 20 percent gap between men and women on the engineering path who eventually earn degrees in engineering, among the most qualified students there is no difference in degree completion rates.

The Empirical Core Curriculum.

● Changes in the empirical core curriculum of engineering students over two decades reflect increases in sub-field concentrations in mechanical and computer engineering and declines in civil and chemical engineering.

● No matter what one's final destination on the engineering path—threshold, migrant, or completer—bachelor's degree recipients spent more time in calculus than any other course. For degree completers in engineering, one out of every seven credits earned was in mathematics.

● Of the groups on the engineering path, the migrants have much higher course participation rates than others in physics, computer science, computer programming, and philosophy, providing some clues as to where these students go when they leave engineering.

● Among engineering degree completers, only four courses outside the sciences, mathematics, and technology—introduction to economics, English composition, general psychology, and introduction to management—turn up frequently on transcripts.

High School Backgrounds.

● The highest level of mathematics studied in secondary school is strongly correlated with bachelor's degree completion in any field. The correlation is stronger for men than women, and stronger, still, for students from the lowest socioeconomic status (SES) quintile. But once students reach the threshold of the engineering path, these effects diminish.

• In terms of high school mathematics and science backgrounds, women and men who come to the engineering path look remarkably alike, yet very different from the women and men who never attempt to major in engineering. Women, however, have a higher academic performance profile (academic grade point average, class rank) than men, regardless of where they end up in college.

• Women who eventually completed engineering degrees had slightly higher SAT scores than male completers and were more uniform in test performance, whereas women who left engineering performed much worse than men on the SAT and evidenced greater variance in performance.

• About 4 percent of high school graduates with curricular momentum in mathematics and science and high quality academic profiles were not interested at all in engineering, rather, for the most part, in "pre-professional" preparation in college and (for women) in health sciences/services majors. Women constitute 60 percent of this high-talent group, and among high-talent students, very few input measures can be squeezed to explain nuances in subsequent student choice.

Choice and Attrition in Engineering.

• As evidenced among labor market participants at age 28/29, engineering attracts a high proportion of people who had a consistent occupational goal starting in high school and a low proportion of people who were constantly changing their career objectives.

• Women who intended to major in engineering enjoyed the highest degree of parental support for bachelor's degree attainment among all women—or men—who intended to major in any field.

• Once in higher education, there is considerable "traffic" among the disciplines, some (but not all) of which can be explained by students' curricular momentum. Students who migrated from the engineering path did so primarily to disciplines requiring strong quantitative skills—computer science, business, and physical sciences—skills in which these students had made considerable investments.

• Credit loads in engineering are not much higher than those in other fields, though engineering students perceive overload because of a high ratio of classroom, laboratory, and study hours to credits awarded. The perception of overload is one of the major factors involved in decisions to leave engineering.

• Women and men earn similar grades in engineering courses; and the women who leave engineering have higher grades than the men who leave. Women who leave engineering do not leave because of poor academic performance, though they do evidence a higher degree of academic dissatisfaction.

Selected Major Themes

(1) Curricular momentum begins in secondary school, and sets up both trajectories and boundaries. Secondary school mathematics study is the key booster to these trajectories, with performance in trigonometry the gate to potential science or engineering majors in college. The trajectories accelerate and the boundaries become more defined in college. Curricular momentum explains why nearly half the students who leave engineering (the migrants) eventually earn bachelor's degrees in the physical sciences and computer science.

(2) There are considerable differences between engineering and science that confuse students in high school and eventually come into play in field migration. Engineering practice, as students discover only in time, involves clients (and all the ambiguities, cultural contexts, and negotiations that come with clients) far more than the practice of science, and client specifications lie at the core of engineering design. The differences in the culture and texture of engineering and science are highlighted in women's experience in both the college laboratory and the workplace.

(3) The metaphor of "paths" is a far more flexible and accurate way to describe student histories than "pipelines." We cannot micromanage choice, and judge a system to be deficient because students are constantly exploring, acquiring, and changing academic identity. "Pipelines" with "leaks" are convenient metaphors of institutional policy, but they neglect both the texture of student histories and the nature of the paths students discover, sometimes with many detours. What we can do is to improve the signs along the pathways, and, in the case of women in engineering, improve the quality of instruction and professorial sensitivity to women's minority status.

Conclusions

This monograph concludes with a number of suggestions for changing the image of engineering among high school students and potential college majors, particularly women. Given what we know of actual practices in different kinds of engineering workplaces, whatever negative views students have ought to be reexamined. There is just as much complexity and difference, joy and difficulty in the engineering workplace as there is in other occupations. Engineers are not a monolithic gang of boys "tinkering" in a technological "sandbox," and telling bad jokes about incompetence. Foremost among the suggestions is that neither women nor men will choose engineering for the right reasons unless the profession can reach out to a broad population with a full portrait of the richness of its culture and practice, and with a clear map of its intersections with and divergences from bench science.

The study also concludes with suggestions to other disciplines for undertaking similar tracking studies, particularly in fields such as psychology or nursing, where men have been a distinct minority.

Introduction:
Metaphors of Passage and Participation

One of the most persistent questions in higher education concerns the rate and fate of student progress toward credentials. Provosts, deans, and state legislators naturally seek evidence that students entering college progress toward and complete degrees—and within a reasonable amount of time. The terms of measuring progress include "retention," "persistence," and "attainment." But department chairs and the directorates of professional and learned societies have another question, one that we might label "field attrition" or, better, "field migration." Simply put, do students who begin to specialize in a given field in college complete a degree in the same field, and if not, to what fields do they migrate and why? The question is also important to provosts, deans, and state legislators. It bears on academic planning and staffing, on what is known in the trade as "enrollment mix," and on the allocation of non-instructional resources. If we know the comparative shares of total undergraduate enrollments by major, and the way these shares are likely to shift as students move through college, we can budget, make capital investments, and staff with greater efficiency. A reflexive information system within an institution can also provide guidance both by analyzing the characteristics of students who major in different fields, and by seeking feedback from students themselves concerning initial and (if applicable) subsequent selection of major field.

It's not that the choices of entering college students are set in stone (Astin, 1977; Pascarella and Terenzini, 1991; Astin, 1993). And a modest proportion of students enter college undecided as to their major, though this feature of student choice seems to be modestly correlated with institutional selectivity.[1] Academic administrators are of two minds about this: they prefer certainty in planning, but they must respect the process of learning and growth within which students discover their own preferences. Rare is the high school that can introduce a student to linguistics, for example, and very few high school students understand how the study of chemical engineering differs from the study of chemistry. We expect students to explore such possibilities and distinctions in the course of their college careers, and we expect them to develop an academic identity.

As one college dean hyperbolized, "if that [developing an academic identity] means that I process more change-of-major forms a year than we have students, then that's what it means!" The dean was swift to add, though, that "I wish our pre-college radar screen had better electronics and that our academic advisers had—and used—three-dimensional information." When inadequately-informed advisement and the search for academic identity intersect, the results both drag down degree completion rates and stretch time-to-degree, the two basic outcomes of higher education that are of most concern to provosts, deans, and state legislators, let alone students themselves. Some 8 percent of students entering higher education directly from high school wind up, at age 30, with more than 60 credits but no degree whatsoever, partly as a consequence of mediocre performance, but also because they wander from major to major, and with each change of major, backtrack to pick up a course or two to meet requirements, and postpone—perhaps forever—the completion of an academic identity.

1

Student "traffic" among the disciplines (Astin and Astin, 1993) moves at an even higher rate. In the national sample we will use in this monograph, 42 percent of those students who indicated an intended major for their college careers and who earned bachelor's degrees by age 30 actually earned the degree in their intended field. Table 1 indicates the comparative extent of this phenomenon by general major field and for men and women:

Table 1.—Percent of 4-year college students who completed bachelor's degrees by age 30, by sex and intended field as indicated in grade 12

	Proportion Who Earned Bachelor's		Proportion Who Earned Bachelor's in Intended Field		Proportion of Those Indicating Intended Major	
	Men	Women	Men	Women	Men	Women
ALL:	67.2	66.2	43.7*	40.8*	100.0	100.0
Intended Field						
Life Sciences	77.6	79.7	47.3	58.2	3.2	3.0
Computer Sci/Math	70.6	59.6	38.5	32.7	11.0*	6.9*
Engineering/Archit	68.8	77.1	54.3*	21.3*	22.8*	5.4*
Physical Sciences	85.8*	60.0*	33.4*	10.9*	4.8	1.8
Health Sci/Services	77.0*	64.4*	27.6*	50.4*	2.5*	11.7*
Business	68.7	63.6	71.8	63.7	20.2	19.2
Education	52.4*	70.5*	16.9*	57.8*	2.0*	7.2*
Applied Social Sci	66.7	64.8	57.4	46.0	3.8*	7.5*
Social Sciences	64.1*	74.4*	54.4	49.2	5.2*	9.5*
Humanities	74.3	68.8	44.7	44.9	2.7	5.6
Fine/Perform Arts	48.6	55.5	65.5	57.1	2.1	4.0
"Pre-Professional"	72.2	75.7	N.A.	N.A.	11.7	10.0
Other Fields	56.1	64.6	N.A.	N.A.	5.1	2.2
Undecided	51.9	59.5	N.A.	N.A.	3.1*	6.1*

NOTES: (1) "Applied Social Science" includes communications, public administration, social work, home economics (though nutrition/dietetics degrees are included with health sciences/services). (2) N.A. = Not Applicable. (3) *Differences between men and women are significant at p≤.05. (4) Universe consists of all students who indicated an intended major in grade 12, for whom a 4-year college was the true institution of first attendance, and who subsequently earned more than 10 credits from a 4-year college. Weighted N=1.08 million.
SOURCE: National Center for Education Statistics: High School & Beyond/Sophomores.

The reader will immediately observe that the overall bachelor's degree completion rate for this national cohort seems high (two out of three) in light of what one usually reads in the newspapers, but that is a function of the way the universe is defined and the long-term (11 years from high school graduation: 1982–1993) nature of the study. This degree completion rate has not changed in 20 years (Smith *et al*, 1996, p. 25).

Intended major, of course, is not equivalent to the actual first declaration of major (and the "pre-professionals" and undecideds eventually tell the deans what they are going to do). But it is important here to acknowledge that the overall proportion of 42 percent completing degrees in their intended field is a condition of our existence in higher education. This rate may vary from one entering class to another, and vary, too, on the basis of when one asks the question about intended major (see part 5 below). For students entering higher education at the traditional age, as this study hopes to demonstrate, the migration rate is a by-product of factors in student choice of field including (but not limited to) curricular momentum and quality of academic performance carried forward from high school and/or established early in college careers, the growing trend toward multi-institutional attendance, the nature of community college curricula for transfer students, credit loads and stop-out behavior, classroom experiences, changing student perceptions of the labor market, and student misconceptions of what given fields of study and occupations are all about.

Why Engineering as a Case?

Engineering is a discipline that can illustrate the features of student choice that affect field migration and attrition in very clear terms, and was chosen for this study because, however complex its story, all the variables affecting choice, persistence, and migration come into play. Because undergraduate engineering programs are offered in a limited number of institutions, we can offer a sharper primary story line about student history and choice. This sharper story will provide deans in search of "three-dimensional information" with better radar-screen electronics.

Engineering was selected for another reason, one that structures the second story line in this study. For some fields, student choice is particularly important on the grounds of what one might call "equity policy." That is, where, historically, there has been a severe imbalance in participation by gender and/or race, the affected disciplines have made special efforts over the past two decades to recruit and retain students from underrepresented population groups. Since success is measured in proportions of students majoring in a field, this competition among disciplines takes place in a finite glass: the proportions always add to 100 percent, and where there are disciplinary "winners" there must be disciplinary "losers." Initial success in recruitment, too, does not automatically translate into success in retention (Moller-Wong and Eide, 1997; Grandy, 1994; LeBold and Ward, 1988).

Field attrition in a discipline with historical equity problems is not a happy situation. It is particularly unhappy in professional fields of study that lead to licensure. The profession itself, considered as a labor force, comes to exhibit what some economists call

3

"segmentation" (England, 1984). That is, individuals from certain demographic groups are found both concentrated and dominant in the occupation. While a degree of segmentation in the professions may be beyond the control of higher education, excessive concentrations such as those found among nurses, elementary school teachers, and engineers is worrisome. For any human service economy to work efficiently, the specialization of labor should be based not only on learned skills, acquired knowledge, and developed talent, but ability to communicate effectively with a demographically diverse group of clients. There is, in fact, an economic utility of more demographically balanced work forces.

When one compares the proportion of men and women who earned bachelor's in their intended field (table 1), one finds four general fields that evidence a combination of low intention to major *and* low field completion. Two of these fields, education and health sciences/services, affect men. The other two, physical sciences and engineering/architecture, affect women.

The reasons for choosing engineering in examining the gender issues in field migration are evident in table 2. Taking account of the changing gender mix and growth rates of undergraduate education over the past 15 years, table 2 displays what Alsalam and Rogers (1991) termed a "female field concentration ratio"* that uses three temporal reference points:

- the year prior to the modal year of college entry for the cohort that provides the data for our story, 1980–81;

- the modal year of bachelor's degree attainment for that cohort, 1986–87; and

- the most recent year for which these data are available, 1993–94.

During the period 1980–1994, the number of bachelor's degrees awarded annually increased by 200,000; women's share of those degrees increased by five percent to a solid majority; and, most importantly, the field distribution of those degrees changed. The proportion of all degrees that were awarded in computer science, for example, declined dramatically from 1986–87 to 1993–94; the proportion of business degrees rose substantially between 1980–81 and 1986–87; architecture's share of degrees shrank slowly during the period while that of engineering rose. These swirling currents of choice and academic fashion make definitive statements concerning trends difficult.

---

* The proportion of all women who earned bachelor's degrees who majored in a specific field divided by the proportion of all men who earned bachelor's degrees who majored in the same field. Changes in the ratio indicate whether field differences between men and women are widening or contracting, and thus provide some hints of who will be entering the labor market in different occupational areas. This analytic tool is different from "segregation indices" and the segregation curves that are independent of the fraction of women in the analysis (Ransom, 1990).

4

Table 2.—Female field concentration ratios at the bachelor's degree level, 1980–1994

	1980–1981	1986–1987	1993–1994
Engineering	.12	.15	.16
Physical Sciences	.33	.37	.42
Computer Sciences	.49	.50	.33
Architecture	.40	.56	.46
Mathematics	.75	.84	.72
Social Sciences & History	.80	.74	.72
Life Sciences	.80	.90	.88
Business/Accounting	.59	.82	.95
Communications	1.22	1.41	1.20
Visual & Performing Arts	1.77	1.52	1.26
English Language & Lit	1.50	1.81	1.61
Psychology	1.88	2.09	2.27
Education	3.03	3.01	2.84
Health Professions	5.18	5.57	3.92

SOURCES: *Digest of Education Statistics, 1987* (table 159); *Digest of Education Statistics, 1989* (table 215); *Digest of Education Statistics, 1996* (table 260).

We can feel confident, though, that since the mid 1980s—and with the exception of psychology—there has been a moderation in the gender segmentation of traditional female disciplines. This trend is accounted for, in part, by the numbers of women moving into business fields, particularly accounting (Flynn, Leeth, and Levy, 1996); and business has become the most gender-neutral of undergraduate majors. We can also say that engineering has remained consistently at the bottom rank of female representation, not only in the United States (Dorato and Abdallah, 1993; Organization for Economic Cooperation and Development [OECD], 1997[2]), and that there has been only modest improvement in the relative representation of women in the field.

What Is This Study About?

This monograph is, foremost, about the paths taken by college students who reach the "curricular threshold" of engineering. It asks who crosses that threshold, what they brought to the threshold from secondary school, how they subsequently perform, and what happens to them. While I will note and offer some observations about the occupational destinations of our subjects, by "what happens to them," is meant, principally, educational experiences and attainment by age 30. These experiences include details on the content of their study, number and types of institutions attended, degrees earned and ultimate major field. The model can be used in undergraduate majors where curricular thresholds—that is, common patterns of coursework across many institutions—can be identified from student records. In the process, and with the assistance of a remarkable national data source, I hope to help redefine what we mean by "field attrition," and to turn some conventional wisdoms concerning who engineering students are and what happens to them into mythologies. The model of engineering should throw into bold relief the types of information that might be gathered at the institutional level to facilitate a more realistic assessment of student choice and progress.

Secondly and critically, the monograph is about the differential paths traversed by women and men as they moved toward the engineering path from their high school days, some backing off before arriving in college, some testing the terrain, some crossing the threshold but then migrating to other fields, and some actually completing degrees in engineering. The emphases of this story are those that emerge from the generic tracking of engineering students, hence they do not cover all the variations from the literature on women's education.

The analysis treats the generic account first. The gender story does not begin until part 4, then becomes a stronger line as the analysis progresses. Both stories are driven by historical method and objectives: to come as close to a true tale as the evidence allows. The remarkable evidence, in this case, has never been used before for this purpose.

Data Source and Its Limitations

Our data come from a national age-cohort longitudinal study, and rely heavily on the college transcripts of participants in that study. The study, conducted over 13 years by the National Center for Education Statistics (hereafter referred to as NCES), followed the high school graduating class of 1982, known as the High School & Beyond/Sophomore Cohort (hereafter referred to as HS&B/So). The college transcripts were gathered between February and September of 1993, when the members of this cohort were 29/30 years old. Labor market histories are complete through mid-1992, when the last of the surveys of this cohort were conducted. The HS&B/So, like its predecessor ten years earlier (the National Longitudinal Study of the High School Class of 1972) and its still-in-progress successor (the National Education Longitudinal Study of 1988), is an incredibly rich data archive, and includes high school records, test scores, and data on family background, family formation, changing attitudes and opinions, financial aid files, and military service, for example. The sample is robust: 14,825, of whom we have high school records for 13,020, test scores for 12,969, postsecondary transcripts for 8,395, labor market histories for 12,640. No other national longitudinal studies datasets have all this information—and at such high response rates[3]—particularly the information from unobtrusive sources such as high school and college transcripts, and covering such a long period of generational history (from the earliest high school transcript entry to the most recent graduate school transcript entry can be a span of 15 years). Its power as an analytical tool will become apparent in the course of this analysis.

However robust the sample as assembled in the 10th grade, the subsequent life-courses of students have a winnowing effect on the range of analyses. When it comes to describing the careers of students who crossed the curricular threshold of engineering studies in college, the size of the sample is more modest. The N for this group is insufficient for analyses by either

6

race or engineering sub-field, and often yields large standard errors of estimates (see Technical Appendix). And because of the low participation rates of women in engineering, we must aggregate the categories of academic career histories in order to produce statistically significant comparisons[4].

How good is the HS&B/So sample compared to the actual national census? In the matter of engineering degree completers, the average annual census for 1987–1991 was 64,800 (Heckel, 1995); the weighted HS&B/So engineering degree completers group for the over-lapping period, 1986–1993, was 63,736. That is a stunningly close match!

National data sets that allow tracking of student careers have a serious limitation: they do not provide the kind of information that allows direct judgment of the quality of student learning. Thus this monograph cannot ask whether a generation of engineering students mastered design problems or theoretical constructs, or how much the non-engineering portions of their curricula influenced their ability to set engineering problems in larger contexts. We do not know, for example, how well they understand viscous flow or how well they can apply that theoretical knowledge to a practical problem involving ferroliquids on the sealant rings used in space shuttles. We can—and will—make some estimates of the general contours of undergraduate engineering curricula as experienced by students and of the relative strengths of engineering sub-fields in the valises of knowledge with which engineering graduates leave college. But that is as far as one can reach toward evaluating the quality of student learning with a transcript sample from students who attended over 2,500 institutions.

Institutional, Elite, and National Cohorts

Historically, engineering has been a very self-conscious discipline, one in which education "has been one of the most studied activities" (California Postsecondary Education Commission, 1981, p. xi). As is the case in other licensure fields, an enormous amount of literature is devoted to instruction, curriculum and assessment. The teaching interests of the field are not tucked away in the back sections of scholarly journals: there are separate journals, and electronic bulletin boards on which engineering faculty exchange notes on experiments, assignments, classroom processes, and test items. These existed long before the Internet (Warren, 1989).

In many institutions, students are admitted to the engineering program and the college simultaneously, and the literature on engineering education evidences considerable concern with field attrition, particularly among women and minorities. One body of the literature focuses on psychological profiles of students in relation to academic performance, degree of commitment to the field, learning styles, and such cognitive traits as spatial ability (for example, Greenfield, Holloway and Remus, 1982; Tobias, 1990; Hackett, Betz, Casas and Rocha-Singh, 1992; Peters, Chisholm, and Laeng, 1995; Seymour and Hewitt, 1997). Another important research line carries a more "economic" tone, in that it takes the engineering profession (or engineering as a sub-species of the practice of science and technology) as its setting, and considers organizational culture, modes of work, and occupational dynamics in relation to the experience of underrepresented groups (Saigal, 1987; Brush, 1991; Ellis and Eng, 1991; McIlwee and Robinson, 1992; Etzkowitz *et al*, 1994; Traunter, Chou, Yates, and Stalnaker, 1996). While this literature is not always explicit in the matter, there is no doubt that messages of the experience of women, in particular, filter back through the educational system to influence student perceptions, behaviors and decisions (Morgan, 1992; Didion, 1993; Henes, Bland, Darby and McDonald, 1995).

With few exceptions (Astin and Astin, 1993; California Postsecondary Education Commission, 1986; Grandy, 1995; Seymour and Hewitt, 1997), the studies that relate student psychological profiles and teaching modalities to retention and completion in the field are individual institutional research efforts and the institutions at issue are usually four-year colleges. A cohort of students who declare engineering as a major upon entrance to college is identified and tracked from the introductory engineering design course onward (e.g. Schonberger, 1990; Epstein, 1991; Humphreys and Freeland, 1992; Ginorio, Brown, Henderson and Cook, 1993). Given the restricted nature of some of the populations, the conclusions of such studies may be more suggestive than generalizable.

A fine example within the universe of these studies began with 124 students in an introductory (sophomore level) chemical engineering course at North Carolina State University in 1990 (Felder *et al*, 1993; Felder, Mohr, Dietz, and Baker-Ward, 1994; Felder *et al*, 1995; and Felder, 1995), and sought to explain their academic fates (completion, retention, GPAs) with multivariate statistics controlling for sequences of background variables, initial experiences in the engineering curriculum, and responses to a five-semester experimental course sequence. N.C. State is the flagship engineering school in the North Carolina higher education system, and its engineering programs are selective. Felder and his colleagues were interested in psychological profiles and correlates of progress in student behaviors ranging from extracurricular activities to test strategies. Grades and expectations about grades/performance play significant roles in the analysis. We emerge with a rich sense of the students, instructional methods, and constructive suggestions for classroom and course-sequence design. The demographics of this group of students, though, seems to limit the generalizability of the study: nearly half of them came from rural high schools, twice the proportion in the national sample used in this monograph[5], and were not as well prepared as their urban and suburban peers (Felder, Mohr, Dietz and Baker-Ward, 1994).

Another line of retention studies broadens the population to all science, mathematics, engineering and technology (SMET) students, but then restricts the population to elites, thus censoring the behavior of the mass. Seymour and Hewitt (1997) for example, interviewed 335 students with SATQ scores of 650 or higher; Strenta *et al* (1993) confined their study to highly selective institutions; Tobias' (1990) "second tier" subjects were all graduates of elite schools. Some 95 percent of U.S. undergraduates do *not* attend elite schools or score above 650 on the SATQ. Lessons from the right tail of a distribution have rarely been generalizable, though they might provide some hints of what to look for elsewhere. For example, if you start with a group of students whose SATQ = >650, common sense says it is hardly likely that they will leave a field because of "personal inadequacy in the face of academic challenge" (Seymour and Hewitt, p. 32). If that is true for students with a high general learned ability quotient, it comes to serve as a benchmark for other hypotheses that may apply to a broader population, for example, students' growing understanding of the nature of work in different occupations in relation to their own proclivities.

The mini-longitudinal studies involving institutional cohorts are very helpful. But they would teach us more if they were set against a national tapestry of potential engineering students, including those who transfer from two-year to four-year colleges. National samples of college transcripts from the NCES longitudinal studies (as well as the surveys of the Engineering Manpower Commission[6]) teach us that not all engineering majors declared engineering as a major on entrance to college, that some of those who did declare engineering never followed up on their declaration, and that a better approach to identifying potential engineering majors is to find those students who explore the lower boundaries of the engineering curriculum and its co-requisites to reach a "threshold" of the field. These students have demonstrated sufficient interest and (in some cases) achievement to cross the threshold, and pursue further work in engineering. As we will see, many students who reach the threshold do not cross. In fact, some of them simply happen across the threshold in the course of other lower division undergraduate work, and have no intention of following the path towards an engineering degree. Others may attend technology institutes with core freshman year curricula that cover the threshold course work, or the U.S. military service academies that require a minimum number of engineering courses. For these students, this accidental "threshold" coursework may not be concentrated in the first year of study. They are nonetheless included in the analysis because they share enough of a curricular experience with others on the engineering path that they could return to these byways, with purpose.

The HS&B/So database and the stories it reveals differentiate this exposition from those based on the intermittent longitudinal follow-ups to the national entering freshman classes surveyed by the Cooperative Institutional Research Project (CIRP) of the Higher Education Research Institute at UCLA. CIRP has conducted annual surveys of entering college freshmen since 1966. In 1982, for example, when most of the HS&B/So students entered college, CIRP surveys from nearly 190,000 students from 350 institutions were included in the national norms (Astin, Hemond, and Richardson, 1982). The universe is stratified by institutional control, size and selectivity, and student data are weighted differentially within stratification cells. The result is representative of first-time freshmen, though part-time students and transfers are excluded from the national norms. The survey collects a considerable amount of information from a very large number of students about pre-college activities, secondary school curriculum and performance, educational and occupational aspirations, values and attitudes. As a consistent series of cross-sectional portraits, the CIRP freshman surveys present reliable and intriguing trends.

Our principal interest in the CIRP data, though, lies in three occasions on which longitudinal follow-ups to a given entering class were conducted, as these have formed the basis for an enormous amount of research on the impact of higher education on students. The grandmother of these occasions was a 1972 follow-up to the entering freshman class of 1968, the data from which formed the basis of Astin's seminal *Four Critical Years* (1977). The occasions of the greatest interest to our study are a 1989 follow-up to the entering freshman class of 1985 (used in Astin, 1993; and Astin and Astin, 1993), and a 1991 follow-up to the entering class of 1987, the unpublished data from which were provided to Seymour and Hewitt (1997). This literature contains a variety of conclusions about engineering students and their experience that I attempt to replicate using a different national database and a different approach to defining the population of "engineering students." It will come as no surprise that, in matters of curriculum, performance and academic attainment, a database grounded in high school and college transcripts (including all institutions attended by the student) will agree but occasionally with a database grounded principally in surveys and with interests confined to the institution in which the student was first surveyed. This contrast also applies to the "National Graduation Rate Study" of 204,000 entering freshmen in 38 public, land grant, research universities in 1988 and 1990 (Kroc, Howard, Hull, and Woodard, 1997) that we will also have occasion to cite.

Metaphors of Passage: What Is a "Path"?

For decades, the higher education literature has struggled toward a language that reflects student careers, particularly in scientific fields. The problem with the most common of the metaphors, e.g. "pipeline," is that they are driven by frameworks of professions, industries, work forces, and/or institutions, not individual student behavior. What students do, after all, cannot be described very well by "pipelines" with "leaks." The metaphors are children of policy needs, not helpful descriptors. Think about the difference in values inherent in the verbs "leak" and "migrate." In the context of human resource development, the subject of the former is a fluid mechanics model and that of the latter a sentient human being.

A "path" is a story-line created by a central actor, in this case, a student. It is not a paved roadway with exit ramps at set intervals, rather a trail that one constructs along contours of the terrain. One can wander away from a rough trail marked by the footsteps of predecessors, finding another pathway that may fit one's proclivities and changing values better. One may detour and return, and, in the detour, establish an alternative way to get there from here. And "there" is not necessarily an immutable, fixed place. A path through higher education, after all, is not merely one of curriculum. It is also very much about student growth, the discovery of interest, the sanding down of sharp edges, the construction of refuges, the honing of negotiating skills, and the development of behaviors and stances to serve in the workplace, family formation, and community life.

The word, "pathways" is more common than "path" in the science/mathematics/engineering/ technology (SMET) education literature. But the term is invoked almost reflexively, and its meaning is not always clear. The Howard Hughes Medical Institute's report of its 1996 undergraduate program directors meeting, *Assessing Science Pathways* (1997), for example, includes such notions as "tracking the test scores and course choices of students who have participated in science education programs and research experiences" and "[assessing] changes in attitudes about science . . ." (p. xi).

The "pathways" literature frequently includes accounts of mentorships and research experience as critical to persistence in science, and as setting off chains of experiences that wind up in medical schools, post-docs, or corporate research labs. These stories are frequently—but not always (see, e.g. Zuckerman, Cole, and Bruer, 1991)—uplifting, particularly when they involve women and underrepresented minorities. The student records in NCES longitudinal studies unfortunately do not always show special research experiences, and the surveys have never asked questions about mentorship. These are limitations of databases such as the HS&B/So. On the other side, the "pathways" reports are principally anecdotal and do not detail the "chains of experiences," including high school and college coursework, that are essential to the results. When they do (e.g. Vetter, 1988), they lump all SMET fields together, thus failing to discriminate the critical differences between the culture of bench science and the culture of engineering that emerges from the empirical data of our longitudinal studies and that, we can reasonably hypothesize, plays a role in student choice. Indeed, a major objective of this monograph is to remind people just how much engineering is *not* science, and how confounding the two hampers our ability to advise college students and track their progress (see part 3 below).

In a methodological sense, the concept of "path" used in this monograph is loosely analogous to that used in the structural equations of statistical prediction. That is, while we are examining a temporal sequence over a period of 15 years[7], we start with a "threshold" behavior that occurred roughly one-third of the way through this history, and work out in both temporal directions from this threshold. The threshold censors the population under consideration. A structural equation would focus on a sequence of experiences from year 1 (see, e.g. Grandy, 1995), and sort effects by temporal criteria. In contrast, this study begins with a set of behaviors and performances in the second phase of that temporal sequence, namely, early college experience: what students studied and how well they performed academically. We then look to both the antecedent and subsequent academic history of the students who have been sorted-in by the threshold criterion.

The inquiry of this monograph is not interested in prediction, rather in relationships of experiences. I take this approach because the most convincing statistical methodologies of prediction reduce the richness of experience to dichotomous variables. There is nothing wrong with those techniques, but if one is interested in the detours and variations of human behavior, dichotomous variables (for example, earned a degree/did not earn a degree) are seriously lacking. Thus, too, no causal chains are posited, though now that the longitudinal study has run its course we can make stronger statistical cases for relationships of experiences than were available to earlier students of the HS&B/So cohort in the matter of the SMET "pipeline" (for example, Hilton and Lee, 1988).

Outline of the Exposition

Part 1 performs three tasks critical to the entire story. First, it delineates the empirical "threshold" of the engineering path and the various "destinations" of the HS&B/So cohort on that path by age 29/30; this is the basic analytic framework, derived inductively from transcript records. Second, it indicates the various institutional attendance patterns of this group that render the story somewhat more complex than previous research on engineering field attrition—let alone the curricular history of *any* group of students—has been able to grasp. Lastly, it describes the role of the community college not only in the preparation of transfer students but also in terms of those whose programs are confined to sub-baccalaureate preparation in engineering technologies. The community college portion of the engineering path is not usually acknowledged in the literature.

Part 2 will show precisely what students on the engineering path studied in college, both in terms of proportions of total credits earned and in terms of participation rates in different course clusters. We will compare the "empirical core curriculum" by both engineering path status and by cohort (the HS&B/So engineering graduates of the years 1986–1993 against that of engineering graduates of the years 1976–1984). This section departs from other national science/engineering "pipeline" studies in the attention it pays to content.

Part 3 is an interlude. It reflects on some of the differences between engineering and science, and the culture of engineering education and practice. The purpose of the interlude is to set the stage for deeper analyses of field attrition and the comparative careers of men and women in scientific and engineering fields, as well as to indicate the analytical conundrums that result from the aggregation of engineering with other scientific fields.

Part 4 sets out the secondary school background characteristics of students who reached the threshold of the engineering path, and compares the high school attainments of men and women in terms of highest level of mathematics studied, participation in core science courses, academic performance, and test scores. These are all traditional measures—and predictors—of college attendance, but the dependent variable here is not mere access, rather degree completion. Therefore, the analysis departs from conventional formulas in the research literature. It also advances the importance of "curricular momentum" in sketching the boundaries of academic choice, and concludes by examining students who had enough momentum to carry them onto the engineering path, but chose not to enter.

Part 5 discusses the factors involved in student choice of engineering as both a career and a major field of study in college. This section also looks backwards from labor market status in 1991 to college backgrounds. It examines the timing of choice, the consistency of students' visions of career and major, and the characteristics of those students who are "lost" to engineering between 12th grade and the first semester of college. Other disciplines in which major is closely tied to occupation, for example, journalism, education, music, health sciences, computer science, and applied visual arts—can be subject to similar analyses.

12

Part 6 demonstrates how our data on field migration and degree completion in the HS&B/So compare to those from other major studies of roughly the same period. This is a necessary exercise in light of the whirlwind of data presented to institutional, state system, and national audiences. We find some remarkable agreements from very different databases, but some serious differences in analyses of the "traffic" and "balance of trade" among the disciplines.

Part 7 examines the differential experiences of men and women engineering students in terms of classroom environments and behaviors, credit load, and grades. It reviews some of the observations on the correlates of field attrition offered by other major studies that involved debriefings of students, and melds these into a summary of this particular investigation, its challenges to engineering as a profession and culture, and its generic suggestions for attaining three-dimensional tracking of college students in any field.

This monograph tells a particular kind of story. It does not pretend to confront the panoply of issues facing the conduct and context of engineering education in the United States. While it covers issues of the "student pipeline," curriculum and retention, its data source cannot access questions about faculty, teaching methods, costs, or equipment, let alone graduate education, cooperative assignments, and the continuing education of practicing engineers.

In fact, this monograph is not designed for engineering educators and engineering professional associations alone. It is also for:

- Science educators, scientific societies, and science policy-making bodies that often include engineering under their analytic and programmatic umbrellas;

- Higher education faculty in other disciplines, including non-scientific fields, who I hope will be prodded to investigate the virtues of empirically-based trackings of threshold-crossings and migration in their own areas;

- Counselors and advisers at both secondary school and college levels, whose understanding of such factors in student choice as the image of engineering as an occupation, and the impact of curricular momentum and optimum levels of mathematics to be studied, I trust will be enhanced, particularly for advising women; and

- The provosts, deans, legislators, and journalists who I hope will learn enough to set the local cases with which they deal against the three-dimensional histories of college students from a national tapestry.

Part 1—Engineering Paths as Established by Students: Destinations and Attendance Patterns

The "Threshold" and Beyond

The first task in our story is to delineate the threshold and, for those students who stay on the path and take the next step, the stations, or destinations, at which they arrived by age 30. Appropriate to the analysis of field attrition, each destination can be characterized by highest degree attained and major field (whether or not any degree was attained). The paths to all stations will be described in terms of combinations and types of institutions attended, college academic performance, courses taken, and other features of undergraduate careers. The destinations are thus configured in terms of the empirical characteristics and experiences of the students who arrived at them.

The "engineering path" variable was not determined algorithmically. Instead, it emerged in the course of reading standardized records of 8,215 students in the HS&B/So database, line by line. These records contain the transcripts of all institutions attended by the student through age 29/30[8], and the reading was initially directed at a variety of other issues (accuracy of coding, completeness of record, continuity of enrollment, true date of first attendance, and others). A set of general decision rules for classification of students' programs was hypothesized on the basis of the first thousand records reviewed, and modified as we progressed through the balance. There were two readers, who judged each case separately and then compared their classifications. Where the classifications differed, the case was discussed, occasionally referred to external authority, and the differences resolved. By this inductive process, the threshold of the engineering curriculum was defined. To be judged as having arrived at the threshold, a student had to have earned more than 10 credits from a bachelor's degree-granting institution (though the student may have also attended a two-year college) and completed three courses at any institution during the first four semesters (or six quarters) of his/her academic career: (1) mathematics at a minimum level of pre-calculus; and either (2) both the introductory engineering design course and engineering graphics or (3) either the introductory engineering design course *or* engineering graphics *and* the introductory course in an engineering sub-field (electrical, chemical, etc.). Completion means just that. It says nothing about performance: at this stage of the accounting, a D is worth the same as an A. But the criterion of completion excludes all cases where the record indicates a withdrawal from a course.

The three courses cited were the most common in freshmen engineering curricula (other than physics and chemistry) at the time our cohort entered higher education (Kauffman, 1980), though engineering graphics, in particular, is no longer as common (Hendley, 1997). Despite the appeal of the argument that physics "seems to be an intermediary between theoretical mathematics and practical engineering" (Schonberger, 1990, p. 102) and that the first college physics course has greater "filtering" effects on women than men, I did not use physics and/or chemistry among the "threshold" criteria principally because cases of

14

Advanced Placement are not always identifiable on the transcripts, and in those colleges that grant credit for AP, a student's record may not indicate any college-level chemistry and/or physics.

The three courses cited are also the most common across all subfields of engineering. At one time, statics or a course combining statics and dynamics might have been included in this common core, but in computer, electrical and chemical engineering, a significant proportion of departments did not require a formal course in this area during the period our cohort was in college (Heggen, 1988). It should be mentioned, too, that traditional graphics was an almost universal requirement in associate's degree engineering technology programs at the time (Eisenberg, 1987).

A separate threshold was established for students who spent nearly all of their postsecondary careers in community colleges and/or other sub-baccalaureate schools. For this group, the mathematics requirement was lowered to college algebra or an algebra/trigonometry-based technical mathematics course, and engineering technology courses could be substituted for engineering design and the introductory course in a discrete field of engineering. The destinations for these students are categorized only by the level of degree they earned in any technology field: none, certificate, or associate's. While a small percentage of this group eventually earned bachelor's degrees, these degrees were not in engineering, engineering technologies, or architecture. The "2-year program only" students are used but occasionally in this analysis, and primarily for comparison with community college transfer students who become bachelor's degree candidates on the engineering path.

The old models of engineering education are changing. A decade from now it may not be possible to determine the "threshold" of the engineering path by the formula followed in this analysis. Required sequences of courses in mathematics and theoretical science may give way to action-oriented curricula in which the math and scientific theory are acquired on an ad hoc basis. Some engineering schools already use freshman "foundations" courses that integrate calculus, physics, engineering mechanics, and design (see, e.g. Carr *et al*, 1995), and we will have to learn how to pick these up from transcripts in the next longitudinal studies cohort (the high school graduating class of 1992). But the history of the 1982–1993 cohort can still be written "old-style," so to speak.

Only 9 percent (weighted N=181k) of all students in this cohort, no matter what combination of postsecondary institutions they attended, reached these "thresholds" of the engineering path. It is worth noting that this cohort entered higher education during a period of an upward swing in intention to major in science, mathematics, engineering, and technology fields, and just prior to the peak of that interest (Grandy, 1989; Dey, Astin and Korn, 1991). In engineering, the intents of this cohort coincided with the peak (Grandy, 1989). Declared interest and intent to major, however, provide but general parameters of what will actually happen as students learn what they do not know upon entrance to college.

15

Migrants and Completers

Beyond the threshold[9], performance and withdrawal count. The subsequent destinations along the path account for students who "migrated" from engineering to other fields. That is, they took at least two—and often three or four—courses beyond the threshold before engineering was replaced by another field or the record simply ended. For students who attended four-year colleges, the groups that left engineering were classified in two ways, by performance and by curricular destination: science, mathematics, engineering or technology (SMET) or other fields (non-SMET). The performance variable was simple; if the majority of the student's grades in mathematics, core science, and engineering were "C+" or lower, the student was classified as a mediocre performer; otherwise their performance was judged to be adequate/good. Withdrawals (as distinguished from "drops") and incompletes that were never resolved on the transcript were judged to be poor grades, in the category of "C+ or lower." While one might say the data reflect a self-fulfilling prophecy, the GPAs of the two groups of migrants—low performing and high performing—at the end of their undergraduate careers were 2.23 (SD=.353; s.e.*=.117) and 2.80 (SD=.488; s.e.=.082), respectively.

The timing of "migration," defined by custom as "year of study," is difficult to identify. It involves a drifting away from the major, for which reason I prefer the term to "defectors" (Astin and Astin, 1993) or "switchers" (Seymour and Hewitt, 1997). Events such as "changed major" are usually not recorded on transcripts. When they are recorded, they are found as journal entries made by registrars with date stamps that may bear little relationship to the timing of the actual change. Too, in cases of non-continuous enrollment (one out of six migrants stopped out of college at some point in their careers), enrollment that may embrace many summer terms, part-time enrollment, and histories that include withdrawals and failures, it is difficult to say just where one "year of study" begins and another one ends.

The precise nature of coursework beyond the threshold was not a consideration in the determination of the general destination on the path, migrant or completer. Many students might take "mechanics of materials," for example, since it is common to different specialties; fewer would take advanced mathematics courses such as tensor calculus, which is a rare elective. Beyond the threshold we are dealing with specialties, electives, and departmental strengths, and these express themselves inconsistently in a national sample such as ours. Beyond the threshold, too, we found that, within bachelor's degree-granting institutions, the engineering path is shared, in considerable part, by students from two other disciplines: engineering technologies and architecture. The transcripts of architecture graduates (our definition did *not* include graduates in city, community, or regional planning) show that 75 percent completed more than three engineering courses; and those of engineering technology graduates show 91 percent completing more than three engineering courses. Some, but not all, of these students attained sufficient curricular momentum to convert to an engineering major had they chosen to do so. These two groups are thus included in the

---

*s.e.=standard error of the estimate, here adjusted for design effects. See Technical Appendix.

Table 3.—Destinations of students along the engineering path, 1982–1993

	All	Excluding 2-Year Only
Threshold only	14.1%*	18.3%*
Beyond threshold: low-performing migrants		
Switched to SMET field	0.7	0.9
Switched to non-SMET field	2.3	3.0
Left higher education	3.6	4.6
Beyond threshold: high-performing migrants		
Switched to SMET field	4.2	5.5
Switched to non-SMET field	4.6	6.0
Left higher education	1.2	1.5
Still enrolled in engineering at age 30	2.7	3.5
Bachelor's completers: no grad school		
In Engineering	23.6*	30.7*
In Engineering Technologies	5.4	7.1
In Architecture	1.8	2.4
Continuing graduate students		
Bachelor's in Engineering, continued in SMET field	8.3*	10.8*
Bachelor's in Engineering, continued in non-SMET field	4.4	5.7
Two-year engineering tech program only		
2-Year Program Only: No Degree	6.4	----
2-Year Program Only: Certificate	5.8	----
2-Year Program Only: Associate's	10.8*	----

NOTES: (1) *p≤.05; (2) columns may not add to 100.0% due to rounding.
(3) Weighted Ns: all=181.3k; excluding 2-year only=139.5k.
SOURCE: National Center for Education Statistics, High School & Beyond/
Sophomores.

engineering path analysis. So is a tiny group that completed interdisciplinary bachelor's degrees involving engineering and communications technology, computer science, and industrial management. While these groups constitute a small proportion of the whole, it is important to note that they are included and that, in this respect, our story line is slightly different from others in the literature on engineering education.

If we take the finest gradations of the engineering path, two distributions of our universe are presented in table 3. The first includes all students; the second excludes students whose careers were spent principally (though not exclusively) in 2-year college engineering technology programs. As is immediately apparent, this detailed delineation yields very few statistically significant estimates. One can imagine how small and shaky some of these percentages would become if we divided each category by gender, or the categories of those who did not complete degrees in engineering by the highest degree they did complete or the fields to which they migrated. Thus, for purposes of subsequent analyses, I will aggregate these 16 "destinations" on the engineering path into 3–5 categories depending on the question. The most common aggregation will be tripartite: threshold, migrants, and completers, with students in 2-year only programs analyzed separately and excluding students who were still enrolled in engineering at age 29/30 (see table 4).

Table 4.—Tripartite distribution of students on the engineering path, excluding students in 2-year engineering tech programs and students still enrolled in undergraduate engineering programs at age 29/30

	All	Men	Women	Female Proportion of Destination
Total:				14.6%
Threshold Only:	19.0%	18.3%	22.7%	17.6
	(2.48)	(2.69)	(6.37)	(5.26)
Migrants:	22.3	20.0*	35.4*	23.4
	(2.31)	(2.39)	(7.18)	(5.91)
Completers:	58.8	61.6*	41.9*	10.4
	(3.02)	(3.23)	(7.04)	(2.12)

NOTES: (1) Standard errors of the estimates are in parentheses. (2) Columns may not add to 100.0% due to rounding. (3) *Row comparisons of men and women $p \leq .05$. (4) Weighted N=139.5k.
SOURCE: National Center for Education Statistics: High School & Beyond/Sophomores.

Presenting descriptive statistics that evidence little statistical significance sounds like an oxymoron and is usually not a good idea. We allow it in table 3 only to provide some hints as to the kind of fine relationships that might be obtained with larger databases, using unweighted Ns in multivariate analyses (for example, in the type of appraisals that Astin and his colleagues have undertaken for many years). For example, the categories of permanent drop-outs and bachelor's degree recipients who continue on to graduate school exhibit, *prima facie*, very different characteristics and behaviors from the other groups into which they will be aggregated. Long-term non-completers (that is, those who were still enrolled in bachelor's degree programs in engineering as of the last date on their transcript records) are such a small and highly diverse group (including people working on second bachelor's degrees, students who have recently transferred from 2-year to 4-year colleges after a long stop-out period, and students with incomplete records) that they cannot be aggregated with any other group and it is simply best to leave them out altogether.

How well does the gender distribution match data from other sources for the same period? The Engineering Manpower Commission reported that 16.6 percent of entering freshmen engineering majors in 1982 were women. The figure for 1983 was 17 percent (ASEE, 1986, p. 43). Assuming that reaching the threshold is equivalent to being an "entering freshman engineering major," then our female proportion of 14.6 percent for students who reached the threshold is a little low, but is understandable since the HS&B/So engineering path universe includes students who started in community colleges, and men tend to dominate community college engineering technology programs even more than they dominate 4-year college engineering programs. The total number of entering engineering majors reported by the EMC for 1982 was 115,303; our weighted N for a matching group (fall, 1982 entrants only) using the threshold criteria is 124,621. These differences are not that great when one considers that the HS&B/So numbers are inflated by "accidental" threshold students.

Table 4 also illustrates some of the wisdom of aggregation for univariate analyses of weighted national samples. There is no statistically significant difference in the proportions of women and men whose destination on the engineering path ended at the threshold. But the differences in the fraction of men and women who became migrants and in the proportion who completed degrees are modestly significant ($t=2.06$; $t=2.59$). Without aggregation there is no chance of spotting these relationships.

The Path Confounded by Attendance Patterns

One of the key issues in studying field attrition in engineering is that of institutional effects (Astin, 1977; Astin, 1993). Why are institutional effects particularly important in engineering? Because institutions with large graduate programs tend to enjoy extensive research support from both industry and federal agencies, and the effects of size seep down into the undergraduate program: the proportion of sections being handled by graduate TAs, the proportion of foreign instructors (Vetter, 1988), the culture of hierarchy (Hacker, 1981), the variety and sophistication of available equipment and facilities, and the nature of problems and topics that move from the "scholarly canon" of research into the "pedagogical canon." The National Research Council has used this particular set of characteristics to divide undergraduate engineering programs into two tiers (National Research Council, 1986). Astin's 1993 analysis of what produces persistence toward a career in engineering reminds us of obvious environmental effects. First, there is a limited number of institutions available to would-be engineering students (in 1986, the modal graduation year for students in the HS&B/So, it was 311; Ellis, 1987), and because of the capital investment and level of faculty expertise required to mount an accreditable engineering program (Accreditation Board for Engineering and Technology, 1991), the vast majority of these schools will not be small colleges with cozy environments. Yes, there are a few small colleges with engineering programs, but these, e.g. Rose-Hulman Institute of Technology with an enrollment of 1,500 or so, are special mission institutions.

Table 5.—Number of colleges attended by HS&B/So students as undergraduates, and number of states in which those colleges were located, by engineering path status

	Number of Colleges				> One State
	One	Two	>Two		
TOTAL	46.5% (0.81)	33.9% (0.74)	19.6% (0.62)		21.0%
No engineering path	47.1 (0.85)	33.4 (0.77)	19.5 (0.64)		21.0% (0.76)
Threshold only	35.5 (7.09)	29.1 (6.51)	35.4 (7.52)		24.4 (5.84)
Migrants: left engineering	42.7 (5.97)	38.8 (6.17)	18.5 (4.30)		14.0 (3.67)
Completers: no graduate school	28.7 (3.86)	46.7 (4.41)	24.6 (4.00)		26.7 (3.63)
Completers: graduate school	49.1 (7.33)	40.3 (7.27)	10.6 (3.93)		28.6 (7.76)
2-Year college prog in engineering tech	56.4 (5.54)	34.0 (5.01)	9.6 (3.04)		12.1 (3.36)

NOTES: (1) Universe includes all students who earned more than 10 credits and for whom engineering path status could be determined. Weighted N=1.958M.
(2) Standard errors of the estimates are in parentheses. (3) Rows add to 100.0%
SOURCE: National Center for Education Statistics: High School & Beyond/ Sophomores.

Once a student is in a school with an engineering program, Astin says, there is an environmental peer critical-mass effect, such that the higher the percentage of students majoring in engineering, mathematics or statistics, the more likely are students who begin in engineering to persist in engineering. The threshold for acceleration of field retention, he says, is 25 percent, a very high field share in any school. But if one thinks of "technology schools" such as the Illinois, Massachusetts, California, and Florida Institutes of Technology (IIT, MIT, FIT, and CIT), Rennsalear Polytechnic Institute, Worcester Polytechnic Institute, or colleges in state systems that have been given special status as providers of engineering education (North Carolina State, Virginia Polytechnic Institute), of course field retention will be higher. The culture of technical universities is rather distinct (Hacker, 1981).

Now if only students stayed in the same college from the time they entered higher education until they left, with or without an undergraduate degree (bachelor's or associate's), the institutional effects analysis would be very compelling. But as table 5 makes abundantly clear, students are highly mobile consumers of higher education. In fact, they have become more so over the past quarter century. For the high school class of 1972 followed for 12 years on college transcripts (to 1984), 32 percent of students who earned more than 10 credits also attended more than one school as undergraduates (Adelman, 1994, p.27). For a more recent (1989–1994) cohort study that did not include transcripts, 45 percent of the participants indicated attendance at more than one institution within five years of first entry to college (McCormick, 1997).

For the HS&B/So sample, 53.5 percent attended more than one college, and 40 percent of this group attended schools in more than one state. With the exceptions of permanent (at age 29/30) drop-outs, 2-year college engineering tech students, and bachelor's degree recipients in engineering who continued on to graduate schools, students on the engineering path had a higher multi-institutional attendance rate than others, and those who earned bachelor's degrees tended to cross state lines at a higher rate than others. It is very difficult to judge institutional effects on students who earned terminal bachelor's degrees in engineering if 71 percent of them attended more than one school along the way, 25 percent attended more than two schools, and 54 percent of those who attended more than one school crossed state lines in the process!

If so many attended more than one institution of higher education, what kinds of schools were involved? There are four questions to be asked here: What was the true first institution of attendance? What combination of institutional types was involved? Do these attendance patterns bear any relation to the elapsed time between true date of first attendance and the date a student either received a bachelor's degree or left higher education? For those who attended both community colleges and 4-year schools, how much of their undergraduate time was spent in community colleges? Tables 6, 7, and 8 (in combination with table 5) answer these questions. A few words on the variables used in these tables might be helpful.

21

Table 6.—Type of institution of first attendance for HS&B/So students, by engineering path status

	Doctoral		Compre-hensive		Community College		Other*	
TOTAL	**23.5%**	**(0.81)**	**23.8%**	**(0.79)**	**36.1%**	**(0.92)**	**16.6%**	**(0.63)**
No Engineering Path	21.5	(0.83)	24.6	(0.83)	36.6	(0.96)	17.3	(0.67)
Threshold Only	45.9	(7.61)	14.2	(5.21)	34.3	(7.35)	---	---
Migrants: Left Engineering	50.2	(6.06)	25.8	(5.17)	14.8	(4.06)	---	---
Completers: No Grad School	52.1	(4.34)	17.2	(3.09)	20.1	(3.71)	10.7	(2.64)
Completers: Graduate School	72.9	(5.91)	19.4	(5.28)	---	---	---	---
2-Year Program Only	---	---	---	---	73.3	(4.69)	14.9	(3.53)

NOTES: (1) * Includes liberal arts colleges and specialty institutions such as health science centers, technology schools, and military academies. (2) --Insufficient cases to produce a reliable estimate. (3) Rows will not add to 100 percent due to rounding and low-N cells. (4) Universe consists of all students who earned more than 10 credits and for whom engineering path status could be determined. Weighted N=1.958M

SOURCE: National Center for Education Statistics: High School & Beyond/Sophomores

First, the identification of *true* institution of first attendance is important in the analysis of student careers because some HS&B/So students (a) enrolled in college for credit before they had graduated from high school, (b) sought to get a "jump start" on higher education by taking courses at one institution in the summer following high school graduation and enrolling in a different institution in the fall, and/or (c) evidenced "false starts" or incidental exploratory enrollments. If we called up the first date of attendance in a consolidated student record file, all three of these phenomena would deceive us as to the first school at which a student "made a go of it" following high school graduation. The true institution of first attendance (TRIFA) is that school. Likewise, the true date of first attendance is the first term at the TRIFA, and all elapsed time variables (table 7) are based on that date.

Second, I have aggregated liberal arts colleges with the category of "other" institutions because so few engineering students start out in liberal arts colleges that a separate category would be a statistical wasteland.

Third, in the variable for combinations of institutions attended (table 7), incidental attendance was not counted. For example, if a student enrolled in a 4-year college, and took occasional summer school courses at a community college, that student is judged to be a "4-Year College Only" student. The "2-Year and 4-Year" combination includes not only regular transfers, but also "reverse" transfers (from 4-year to 2-year), and alternating and simultaneous attendance at both 2-year and 4-year colleges.

The numbers that leap off these tables are the following:

- Students who did not reach the threshold of the engineering path at all are a far more diffuse group in terms of true institution of first attendance (table 6), institutional attendance combinations (table 7), and total elapsed undergraduate time (table 8) than students who reached *any* destination on the engineering path.

- One out of three students who reached the threshold of the baccalaureate engineering path but who did not cross the threshold started in a community college (table 6) and nearly 40 percent of this "threshold" group earned more than 10 credits from community colleges (table 7).

- Engineering students who not only complete bachelor's degrees but continue on to graduate or professional school were more likely to start in a doctoral degree granting institution (table 6), enroll only in 4-year colleges (table 7), enroll in only one institution (table 5), and finish degrees faster than others on the path (table 8).

It has been claimed that engineering students have lower rates of enrollment in graduate school (for example, Astin, 1993), but the HS&B/So data show no significant differences in graduate program enrollment rates[10] between engineering degree completers (30.8 percent), migrants (30.2 percent), and others (27.5 percent). These proportions apply to those in the HS&B/So sample who completed bachelor's degrees in 4.5 years in order to match Astin's 4-year (1985–1989) sample. Similar observations have been made of GRE test-takers, even though they are far more diffuse in terms of age and post-baccalaureate job experience (Grandy, 1996).

- One out of five students who received terminal bachelor's degrees in engineering, engineering technology, or architecture (and one out of six students who received bachelor's degrees, terminal or not, in those fields) started in community colleges (table 6); very few of the students who migrated from engineering to other disciplines started in community colleges (table 6) but one out of five used the community college in more than an incidental fashion in the course of their undergraduate careers (table 7).

Table 7.—Combinations of institutions attended by HS&B/So students, by engineering path status

	4-Year Only	2-Year and 4-Year	2-Year Only	Other		Proportion Earning >10 Credits from Community Colls.
TOTAL	**46.5%** (0.97)	**21.2%** (0.69)	**21.6%** (0.74)	**10.7** (0.61)	‖	**42.1%**
No engineering path	45.4 (1.01)	19.2 (0.71)	22.1 (0.78)	13.3 (0.65)	‖	**42.4**
Threshold only	64.4 (7.32)	17.5 (6.51)	10.9 (4.02)	--	‖	**39.4**
Migrants: left engineering	74.6 (5.04)	21.6 (4.68)	--	--	‖	**22.4**
Completers: no graduate school	72.3 (4.06)	26.2 (4.01)	0.0	--	‖	**26.6**
Completers: graduate school	92.1 (2.81)	--	0.0	0.0	‖	**7.9**
2-Year program only	--	17.6 (3.82)	59.0 (5.43)	16.4 (3.83)	‖	**80.5**

NOTES: (1) Universe: all students who earned more than 10 credits, and for whom both engineering path status and combinations of institutions attended as undergraduates could be determined. Weighted N=1.95M. (2) --Insufficient cases to produce a reliable estimate. (3) Rows will not add to 100 percent due to rounding and low-N cells. (4) Standard errors of the estimates are in parentheses.

SOURCE: National Center for Education Statistics: High School & Beyond/Sophomores

Table 8 provides strong evidence of a finding to which we will have many occasions to refer: students who reach the threshold of the engineering path complete bachelor's degrees *in any field* at a much higher rate than those who never reach the threshold. In fact, if we match Astin's 1985–1989 CIRP group or Kroc, Howard, Hull, and Woodard's public research university 1988–1993 and 1990–1995 groups with roughly the same parameters (entered 4-year college, 4.5 years to obtain a degree) our engineering path students have a 10 percent degree completion advantage over those who never reached the threshold (t=3.02), and this spread would be even greater if we moved the "threshold only" students into the non-engineering group. Given the stronger secondary school backgrounds of those who reach the threshold of the engineering path versus those who do not (see part 4 below), this is a common sense conclusion.

Table 8 also helps us confront the enduring proposition that it takes longer to complete degrees in engineering than any other major. Kroc, Howard, Hull and Woodard (1997), for example, make this assertion on the basis of who is still enrolled after five years. The case is not that simple. Consider, first, the mean time-to-bachelor's-degree in table 8. For the entire HS&B/So cohort, it was 4.74 *calendar years*, a modest increase from the 4.51 calendar years for the cohort that graduated from high school a decade earlier. For those who earned degrees in engineering and did *not* continue on to graduate school (at least by age 30), the mean time to degree was modestly longer (5.04 calendar years), no doubt because of the relatively high proportion of this group who attended more than one college (table 5) and/or who transferred in from a community college (table 7). But students who finish engineering degrees *and* continue on to graduate school do so in slightly *less* time than the mean for all bachelor's degree completers. This bi-modal pattern in time-to-degree among engineering completers certainly warrants further investigation.

Community Colleges' Role on the Engineering Path

The most revealing of the data in tables 6, 7, and 8 are those that reference the community college, whose role in the engineering path is insufficiently recognized in the literature. To elaborate on the backgrounds of the migrants, for example, the fraction of community college transfer students was much higher (29 percent) among those who switched to other SMET fields than it was among those who switched to non-SMET fields (10 percent). The community college, then, gave their transfer students enough curricular momentum in SMET fields so that most of them stayed in the territory even if they left engineering. As table 9 reveals, among all students who reached the threshold of the engineering path and attended 4-year colleges, the proportion of community college transfer students who completed bachelor's degrees in any field was almost indistinguishable from the proportion of students completing bachelor's degrees within 4-year college attendance patterns, and the comparative proportions of these two groups completing degrees in engineering is not statistically significant.

How do we know if a community college student is on a trajectory to transfer toward the engineering path? The mathematics courses are a signal as to the type and purpose of the student's program. A course in "technical algebra and trigonometry," for example, is the mathematical foundation for certificate programs in occupational fields such as communications technology (Agrawal and Bingham, 1995), whereas the path toward transfer

Table 8.—Elapsed undergraduate time and time to bachelor's degree for HS&B/So students, by engineering path status

	Total Elapsed Undergraduate Time			Proportion Earning Bachelor's Degree	Time to Bachelor's		
	Time	S.D.	s.e.		Time	S.D.	s.e.
TOTAL	3.94	2.69	.018	44.9%	4.74	1.52	.040
No engineering path	3.86	2.72	.019	43.0	4.71	1.53	.044
Threshold only	4.52	2.07	.130	61.2	4.82	1.26	.249
Migrants: left engineering	4.98	2.18	.127	64.5	4.99	1.81	.368
Completers: no graduate school	5.04	1.40	.059	100.0	5.04	1.40	.277
Completers: graduate school	4.67	1.15	.076	100.0	4.67	1.15	.190
2-Year program in engineering tech	4.28	3.24	.160	N.A.	---	---	---

NOTES: (1) All time is measured in *calendar* years from the true date of first attendance. (2) Standard Deviations (S.D.s) are unadjusted. Standard errors (s.e.) of the estimates are adjusted for design effects. (3) Universe: All students who earned more than 10 undergraduate credits, for whom a true date of first attendance could be determined. Weighted N=1.9M
SOURCE: National Center for Education Statistics: High School & Beyond/ Sophomores.

will show statistics or pre-calculus. For students whose first institution of attendance was the community college and who earned more than 10 credits from community colleges, these distinctions are rather clear in relation to their destinations on the engineering path: 87 percent of those in 2-year technology programs only never studied pre-calculus or calculus, whereas virtually all transfer students who reached at least threshold status had studied pre-calculus or calculus. From the perspective of department heads of engineering technology

Table 9.—Percentage of all students reaching threshold of the engineering path who completed bachelor's degrees by age 30, by transfer status

	Completing Bachelor's		Percent of All
	in Any Field	in Engineering	
Community College Transfer Students	85.1 (4.88)	65.8 (7.39)	**17.8%**
Students in 4-Year Only Attendance Patterns	87.7 (2.08)	60.4 (3.30)	**74.7**
Students in Other Attendance Patterns	23.5 (7.49)	22.4 (7.87)	**7.5**

* Standard errors of the estimates are in parentheses.
<u>SOURCE</u>: National Center for Education Statistics: High School & Beyond/ Sophomores.

programs in 2-year institutions, this bi-modal pattern of preparation results in low program retention rates (Cahalan, Farris, and White, 1990). The HS&B/So data suggest that the attrition occurs, in part, because 17 percent of the students who begin in the 2-year engineering technology programs transfer to 4-year schools (and complete bachelor's degrees) without earning an associate's degree first. For all the complaints of department chairs about both the academic preparation and computer backgrounds of entering students in 2-year programs (see Burton and Celebuski, 1994), the bottom line of "low program retention rates," then, may be due in some measure to talent as well as to the lack of it.

What happens to technology students in 2-year institutions (most of which are community colleges) is determined not only by student preparation but also program orientation. Offering courses—or even credentials—does not necessarily mean articulation with 4-year programs. Given what Burton and Celebuski (1995) report to be the most important objectives of technical education in 2-year colleges—remediation and training in entry-level occupational skills—some certificates and associate's degrees are designed to be terminal credentials. The major programs—in electronics, computer, graphics, and architectural technologies—all emphasize applied skills far more than fundamental science and mathematics (Burton and Celebuski, 1995). Only in two smaller fields—general and chemical engineering technology—does instruction emphasize incontestably transferrable subject matter. And within the broad category of mechanical engineering technology, where the program orientation is heating/ventilating/air conditioning (HVAC) or drafting, even high quality programs will not provide sufficiently transferrable curricula (Eisenberg, 1987).

As table 9 demonstrates, students in non-standard attendance patterns (that is, neither community college transfer nor 4-year college only) have much lower degree completion rates. The non-standard patterns include reverse transfer (4-year to 2-year) and combinations of institutions including proprietary trade schools and other sub-baccalaureate institutions. These students are fairly weak; 59 percent did not pass calculus; and they are overrepresented among the permanent college drop-outs.

So: we have thresholders, migrants, completers, and 2-year program students. These are the major categories of destination. The paths to these destinations have different institutional starting points, and evidence various branches, some of which lead out of higher education altogether, some of which lead through two or more institutions in both traditional modes (transfer from 2-year to 4-year colleges) and non-traditional modes (alternating/simultaneous attendance), some of which involve different periods of residence in community colleges and interrupted periods of enrollment. As we look more closely at these destinations, we inevitably have to talk about what, precisely, these students studied—in college and in high school—that clarifies the terrain through which the paths flow.

Part 2—The Content of Their Curriculum

In particular, our observations on the role of the community college in the travels of students on the engineering path bring curriculum—and the concept of curricular momentum—into the formulas and dynamics. It is very difficult to describe a trail without any sense of its texture, and texture—more than direction—may explain why some choose another path. What must amaze and disappoint the reader of the major studies of the science "pipeline" is how little attention is paid to "real stuff," the content of their curriculum.

What do students at the principal destinations of the engineering path study? In the absence of national collections of syllabi, the transcript histories provide three metrics for answering this question:

    1) absolute time on general content;

    2) proportional time on disciplinary subject matter; and

    3) participation rates in discrete course categories.

Under the first metric, we aggregate credits earned by broad categories such as computer science, physical sciences, social sciences or business. Because credits are a proxy measure for time, this method produces a measure of absolute time on generalized content. Absolute time means little in educational histories because these histories are censored. That is, we end our undergraduate education at a specific point in time, whether or not we have earned a degree, and we can total our credits—our successful endings, so to speak—as of that moment. To say that I earned 12 credits in physical sciences out of a total of 48—with no degree—is very different from saying that I earned 12 credits out of 132 with a bachelor's degree. The *proportion* of my time-on-subject-matter is a measure of the relative weight of knowledge I will take from higher education into the labor market. It is obvious from this example, though, that unless one knows the full amount and content of all credits, this is not a very productive approach.

If absolute time is not a wholly satisfying measure, neither is generalized content. I can earn those 12 physical science credits in astronomy, geology, or general and organic chemistry. In assessing the total configuration of knowledge a cohort of students who start out on the engineering path will acquire, more specificity is necessary. Proportional time on specific content enables us to judge the relative positions and weights of knowledge carried by a cohort. It also enables us to judge the likelihood of a student's curricular path. For example, a student whose first 12 credits in physical sciences are confined to astronomy and physics is not likely to major in biology, but if those 12 credits are in general and organic chemistry, the door to biology is at least open. This common-sense-empiricism lies beyond the reach of multivariate analyses that do not include student records (e.g. Astin and Astin, 1993; Grandy, 1995), and yet is critical to our understanding of the "traffic" among the major disciplinary fields.

Table 10 is an account of proportional time on disciplinary subject matter for students in two cohorts who earned bachelor's degrees in engineering (excluding engineering technology and architecture). There are over 1,000 course categories in the taxonomy used for course coding in NCES's national college transcript samples, *The New College Course Map* (Adelman, 1995). For any one of those categories to account for 1 percent or more of the total undergraduate time for a cohort is a very strong claim for its position in the configuration of knowledge acquired by these students. The 21 course categories are ordered in terms of their ranking in the more recent of the two cohorts (the HS&B/So), account for approximately 60 percent of total undergraduate time of engineering majors, and can justly be called the "empirical core curriculum" of engineering majors. Only three of the 21 course categories lie outside SMET territory, and these categories claim less of the time of the HS&B/So cohort than they did of its predecessor. The degree of concentration in the 18 SMET course categories increased between the two cohorts, suggesting that the depth of study in non-SMET areas by engineering graduates of the mid and late-1980s was shallow.

The way we read this table, for example, is to note that courses in mechanical engineering accounted for 6.9 percent of the total undergraduate time of the 1982–1993 (HS&B/So group), an increase of 2.3 percent over that of the 1972–1984 group. That is a significant increase in the proportion of undergraduate time devoted to mechanical engineering. At the same time, the proportion of total credits accounted for by chemical engineering dropped from 1.9 percent to 1.0 percent. That is a significant decrease. Both trends reflect shifting preferences of sub-specialties between the two cohorts, as do the relative weights of civil engineering versus computer engineering.

Table 10.—Empirical core curriculum of engineering degree completers in two cohorts: 21 courses accounting for the largest percentage of total credits earned

Course Category	1972–1984 Cohort	1982–1993 Cohort	Change
Calculus	8.7	7.1	-1.6
Mechanical Engineering	4.6	6.9	+2.3
Electrical Engineering	8.5	5.9	-2.6
Mechanics, Statics, Dynamics	4.9	5.3	+0.4
General Chemistry	4.4	4.5	+0.1
General Physics	5.3	4.3	-1.0
Pre-Calculus	1.4	3.3	+1.9
Computer Engineering	1.1	2.6	+1.5
Engineering Mathematics	0.7	2.4	+1.7
Civil Engineering	3.6	2.2	-1.4
English Composition	2.2	2.0	-0.2
Materials Engineering	1.2	1.8	+0.6
Intro to Economics	2.0	1.6	-0.4
Engineering: Special Topics	0.9	1.6	+0.7
Electronic Technologies	0.9	1.4	+0.5
Computer Programming	0.6	1.2	+0.6
Intro Engineering/Engin. Design	1.2	1.2	0.0
Engineering Graphics	0.8	1.2	+0.4
Aeronautical Engineering	1.0	1.1	+0.1
General Psychology	1.2	1.1	-0.1
Chemical Engineering	1.9	1.0	-0.9
Total percent of time	**57.1**	**59.7**	**+2.6**

METRIC: Percent of total undergraduate time, using earned credits as a proxy for time.

SOURCES: National Center for Education Statistics: National Longitudinal Study of the High School Class of 1972, and High School & Beyond/Sophomore Cohort.

Other notable changes include the decline in the proportion of time claimed by calculus and the rise of both engineering mathematics (applied calculus and engineering statistics) and pre-calculus. Overall, the proportion of time devoted explicitly to these three mathematics courses rose from 10.8 percent to 12.8 percent between the two cohorts, an empirical trend that reinforces what Hacker (1983) describes as the "mathematization of engineering." When all mathematics courses are included, the HS&B/So engineering students spent nearly one out of every seven credit hours in explicit study of mathematics. The significance of both the proportion of time and the nature of mathematics studied, one can speculate, reflects a bi-modal pattern in the mathematics preparation of students entering engineering programs. About 25 percent of students who reached the threshold of the engineering path had already studied calculus in high school (see table 14), did not need the traditional four semesters of calculus (through differential equations), and could move on to engineering statistics, for example. At the same time, more than half came to the threshold with high school mathematics at trigonometry or less, and inevitably wound up in pre-calculus courses.

One way to validate the influence of sub-field choice on course-taking is to set the weighted N for degrees in the HS&B/So cohort, by engineering subfield, against the actual average annual headcounts reported by the American Society for Engineering Education (ASEE) for the period during which most of the HS&B/So engineering students received their degrees (Heckel, 1995). Using a narrow definition of engineering degrees (that is, architecture and engineering technology are not included) and the six engineering sub-field categories in the taxonomy used for HS&B/So majors, table 11 illustrates the difference between census and sample. The universes are very different, and only for the share of degrees claimed by mechanical and computer engineering, and that in residual fields do the HS&B/So estimates come close to the census. As previously noted, it is remarkable that estimates based on a sample of 10th graders are as close as they are to the census of bachelor's degrees awarded by sub-fields of a discipline a decade later. Here, though, I am emphasizing the fit between degree distribution and proportions of total undergraduate time spent in specific course categories (table 10). The order of sub-field shares of degrees in table 11 matches the order of time-on-content in all engineering courses except in the case of chemical engineering.

If we compare the curriculum taken by the tripartite division of engineering path students, we clearly see what "threshold" and "migration" status mean in terms of the empirical core, and glean some hints as to the disciplines toward which these other groups of students tilted. But in order to present a convincing comparison, we confine the students in table 12 to those who earned bachelor's degrees. In that way, we are measuring proportions of credits earned by the same censoring event. Table 12 also includes, in the universe of engineering completers, those who earned degrees in engineering technology and architecture.

Table 11.—National census versus national cohort estimates of sub-field distribution of engineering degrees, 1987–1991

	ASEE*		HS&B/So	
Total Number of Degrees	64,800*		63,736	
Electrical engineering	20,351	31.4%	15,535	24.4%
Civil engineering	7,050	11.0	3,514	5.5
Chemical engineering	3,640	5.6	8,419	13.2
Mechanical engineering	14,243	22.1	17,696	27.8
Computer engineering	4,018	6.2	5,315	8.3
All other fields	15,498	23.7	13,257	20.8

* Annual average for 1987–1991 as reported by ASEE (Heckel, 1995).

What do we see in table 12? First, no matter what one's destination on the engineering path, bachelor's degree recipients spent more time in calculus than any other course; and that both pre-calculus and post-calculus mathematics (that is, calculus beyond differential equations and such advanced mathematics topics as abstract algebra, combinatorics, and matrix theory) rank in the empirical core of those who did *not* complete engineering degrees, as does engineering mathematics for the completers. Across the entire spread of engineering path destinations, the mathematics content is very strong. The other courses in the "top 20" for the threshold and migrant groups tell us why: the physical sciences, computer science, and accounting are heavily represented, and all are quantitatively based. Later in this analysis, when we note that students who migrate from engineering are more likely to enter these fields than others, it will not be surprising. The quantitative curricular substrate supports student choice in changing fields.

One of the persistent complaints about the undergraduate experience of engineering majors is that the time required for professional preparation ironically does not allow students the chance to develop either the skills for effective client communication or knowledge of culture, law, economics, and ability to maneuver in a global workplace—all of which are *also* part of engineering practice (Board of Engineering Education, 1994). Students get the mathematics, the basic science, the technology, the design courses, the tools courses, and, for practical experience, the co-ops. There is no time, it is said, for anything else. While further training in technical matters and marketing issues occurs on the job (Kunda, 1992; Bucciarelli and Kuhn, 1997), any depth in non-technical knowledge and skills must be acquired off-the-record.

Table 12.—Comparative empirical core curricula for students on the engineering path who completed bachelor's degrees in any field: courses accounting for the largest percentage of total credits earned

Threshold only		Migrants		Completers	
Calculus	4.8	Calculus	6.3	Calculus	6.0
General Chem	3.3	Gen Physics	4.4	Mechan Engin	5.4
Intro Economics	2.9	General Chem	4.3	Electr Engin	4.7
Gen Physics	2.8	English Comp	3.0	Dynamics/Statics	4.3
English Comp	2.6	Electr Engin	2.7	Gen Physics	4.1
Spanish: Lower Lv	2.4	Pre-Calculus	2.5	General Chem	3.9
Naval Science	1.9	Comput Progrmng	2.4	Pre-Calculus	2.9
Pre-Calculus	1.7	Post-Calculus	2.4	Electr Engin Techn	2.4
Intr Computer Sc	1.6	Intro Economics	2.3	English Comp	2.2
U.S. History Surv	1.6	Dynamics/Statics	1.5	Computer Engin	2.1
Gen Psychology	1.6	Intr Accounting	1.5	Architecture Core	2.0
Comput Progrmng	1.5	Adv Accounting	1.4	Engin Mathematics	1.9
Comput Sys Design	1.4	Statistics (Math)	1.2	Civil Engineering	1.7
Adv Accounting	1.4	Organic Chemistry	1.2	Intro Economics	1.6
Intr Accounting	1.4	College Algebra	1.0	Materials Engin	1.5
Statistics (Math)	1.2	General Biology	0.9	Comput Progrmng	1.4
Finance	1.2	Computer Engin	0.9	Comput Engin Tech	1.3
Post-Calculus	1.2	Comput Org/Archit	0.9	Engineering: Other	1.2
Electr Engin Tech	1.1	French: Lower Lev	0.8	Engin Graphics	1.1
College Algebra	1.0	General Psych	0.8	General Psych	1.0
Total % of time	**38.6%**		**42.4%**		**52.7%**

NOTES: (1) The metric is the percent of total undergraduate time, using earned credits as a proxy for time. (2) The universe consists of all students who reached the threshold of the engineering path and subsequently earned a bachelor's degree, whether in engineering or another field. Weighted N=114k.
SOURCE: National Center for Education Statistics: High School & Beyond/Sophomores.

It turns out that this portrait is not wholly accurate, and one of the ways to demonstrate its inadequacy is to examine course participation rates. This approach is very different from examining either absolute or proportional time-on-subject-matter. It asks what proportion of *students* successfully completed at least one course in various curriculum categories. For this calculation we aggregate the 1,000 course categories into 103. Table 13 does this for bachelor's degree recipients across the three major "destinations" groups of the engineering path. It includes the "top 20" course categories in which (with one exception) at least 50 percent of the students in one of those groups of students successfully completed a course. Its basic question is to what extent—and in what curriculum areas—are the participation rates of the engineering completers significantly different from those of the migrants and those who reach only the threshold. Where statistically significant comparisons can be made, the comparatively low participation rates in non-SMET areas for engineering completers appear to be in foreign languages, accounting, philosophy and religious studies, U.S. history surveys, and literature.

At the same time, the migrants evidence considerable strength of participation in physics, computer science, computer programming, and philosophy/religious studies. These features of course-taking provide some clues as to where the migrants go when they leave engineering programs, or, as Astin and Astin (1993) might phrase it, where the "traffic" flows. In those curricular intersections, philosophy fits neatly with physics and computer science.

How do we know whether the results for this sample can be replicated? Are there other benchmarks for the period of the 1980s? A very different and creative approach to these questions of the curricular breadth exposure of engineering majors was taken by the report, *Engineering Education and Licensing in California* (California Postsecondary Education Commission [CPEC], 1981). In a way, this approach would render moot the units of transcript analysis in which we have engaged. The CPEC study analyzed three actual large scale projects from engineering practice to identify 14 "Fields of Understanding" within the history and documentation of those projects. These fields included physical science, ethics, design/application, engineering technology, economics, law, management, engineering science, history, life science, political science, behavioral science, communication arts, and humanities.

Some of these fields translate easily into coursework and transcript entries; some do not. When the 1981 CPEC investigation examined curricular requirements in mechanical engineering at seven universities and set those requirements against the 14 "fields of understanding," it discovered that only two institutions required coursework in more than six of the 14. But requirements do not describe what students actually do, and it is a perpetual mistake of higher education analysts to substitute catalogue statements for empirical evidence of student behavior. Thus, when the 1981 CPEC study examined a small sample of transcripts of graduates from four of these institutions, it found that 85 percent of the students had taken courses in more than six of the 14 "fields of understanding" and that 35 percent had taken courses in more than eight fields (CPEC, 1981, p. 76).

Table 13.—Course participation rates by engineering path destination: percent of students completing at least one course in 20 course categories

Course Category	Threshold	Migrants	Completers
Composition and Writing	92.3*	83.3*	84.8*
Calculus and Advanced Math	78.9*	84.4*	90.0*
General Chemistry	77.6*	79.2*	80.3*
Engineering Sub-Fields	68.1*	79.5*	96.6
Physics (All)	52.9	80.1*	90.4*
Engin Mechanics/Statics/Dynamics	--	49.5	79.3
EnginTech other than Electrical	--	--	73.5
Introductory Economics	72.7*	64.1*	62.6*
General Psychology	72.0	36.0	50.2
College-Level Math+	70.4*	57.6	66.0*
Literature/Letters	68.8	57.1	44.3
Computer Sci (except programming)	62.0*	59.6*	50.6
US History/Amer Civilization	61.8	38.1*	43.4*
Computer Programming	33.7	59.9	49.0
Computer Applications	56.5	38.0*	40.2*
Statistics (Mathematics)	52.7*	44.5*	27.8
Foreign Languages	58.3	43.1	13.4
Engin & Technical Drafting	37.6*	21.8*	55.2
Philosophy & Religious Studies	39.6*	54.8	36.3*
Accounting	47.4*	43.4*	17.0

NOTES: (1) Universe: all students who reached the threshold of the engineering path and subsequently earned a bachelor's degree in engineering or another field. Weighted N=114k. (2) -- The N is insufficient to produce a reliable estimate. (3)* Pairs on these rows are not statistically significant under Bonferroni tests (see Technical Appendix). (4) + Includes college algebra, pre-calculus, and finite and discrete mathematics.
METRIC: Percent of students in each path category who earned any credits in 103 aggregate course categories.

Table 14 matches our HS&B bachelor's degree recipients in engineering against the CPEC findings in "fields of knowledge" outside engineering, mathematics, and physical science[11]:

Table 14.—Proportion of engineering graduates completing courses in broad "fields of understanding" outside engineering

	CPEC	HS&B/So
Economics	82%	67%
Communications	82	90
Behavioral sci	67	54
Management	53	42
Political sci	39	33
Ethics	27	7
Life sciences	23	16
Law	17	8

SOURCES: California Postsecondary Education Commission (1981), p. 78.
National Center for Education Statistics: High School & Beyond/Sophomores.

The differences, of course, reflect the larger and far more diverse sample of institutions in the HS&B/So, the errors that come with the sample, and the earlier time period and state system culture covered by the CPEC data. The discrepancies in the categories of "ethics" and "law" would be narrowed if we assigned courses that are classified as "Science, Technology and Society" (STS) for the HS&B/So cohort to one of these categories, but that assignment would involve an excessive leap of faith. Still, in a licensure-driven profession such as engineering, where the requirements of specialized preparation account for the incredible degree of curricular concentration evidenced in tables 9 and 11, it is hard to imagine that the data in table 13 reflect anything more than the prominence of four or five non-engineering, basic science, or mathematics courses in the curricula of engineering students: introduction to economics, English composition and technical writing, general psychology, and introduction to management. If employers are unhappy with engineering graduates because they have no more than a skipping-stone's familiarity with fields of knowledge that bear on virtually all engineering design problems and interactions with clients, then it is not hard to imagine that some engineering students themselves will be unhappy with the lack of opportunity to acquire sufficient depth in non-technical disciplines, and may drift away from the path toward other fields.

A good example of the depth across fields required of engineering practice: Writing a decade ago, Edward Wenk of the University of Washington wisely saw the advent of "technological delivery systems" as opposed to discrete devices and structures as the products of engineering practice (Wenk, 1988). Engineers are not merely designing, let us say, a cellular telephone, rather the entire system of which the device is only a part. That system, if you think it through, consists of other hardware, software, "socialware," support systems and the training of support workers, telecommunications carrier business practices, and government regulations and tax policies. Engineers do not necessarily need a "course" in each of these areas, though modules on such topics as product safety, client behavior (including behavior in different cultural settings), social and economic forces surrounding that behavior, environmental impact (not merely physical environment, but the social environment as well) can add to the differential perspective required of engineering practice in an age of systems. It is impossible to determine from transcripts whether undergraduate engineering students have been trained in a problem-centered curriculum that includes such features of substance. To say that they have taken a history course or a psychology course or two does not necessarily translate into systems thinking and facility with holistic design.

Part 3—Engineering and Science: Confusing Signs Along the Path

Wenk's example indirectly raises an issue in which our provosts, deans and department chairs must become more fluent in order to achieve optimal fit between student and major: the ways in which the textures of engineering differ from those of science. For the distinctions between engineering practice and the practice of scientific occupations flows from the disciplines themselves and their delivery in the undergraduate curriculum. Herein lies one of our major themes: however closely related, engineering is not science, much of the "traffic" we observe in undergraduate behavior (course taking, change of major) is born of a false conflation of the two, and most of the literature pays inadequate attention to the distinction. Let us spend at least a few moments on those portions of the paths where students encounter confusing directional signs.

Engineering as Enigma

We have novels and television shows about the practice of law and medicine, dramas and motion pictures in which scientists play major roles, and eminently readable memoirs and manifestos of architects. But engineering has no such exposure in mass culture and "is an enigma to the lay public." (Harrier, 1996) The newspapers carry articles about "scientific literacy" and define "practical science literacy" in stories such as breast-feeding versus bottle-feeding babies in countries where ground water is contaminated (Shen, 1975). There is no corollary in "engineering literacy," even when a public safety issue such as seismic design is on the table.

At best, engineering is confused with science in the public consciousness, and that, one can hypothesize, is a noteworthy factor in field attrition. On the basis of their secondary school experience and their command of mathematics, high school graduates have a modest understanding of bench science and its variations in the balance of empirical/theoretical, inductive/deductive approaches in biology versus chemistry, for example. But they have no idea of the nature of engineering until they complete a freshman design course—and even that may not tell them enough. As an indication of how little secondary school students understand what it means to be an engineer, McIlwee and Robinson (1992) note that even when the fathers of their subjects were engineers themselves, that "meant only that the term 'engineer' was not alien to them. It did not mean that they knew what engineers did . . ." (p. 33) At best, high school students think that their experience of mechanics experiments in physics is a proxy for the work of engineers, and, at least among men, this experience influences their decision to major in engineering (Jagacinski and LeBold, 1981).

The students are not wholly wrong, of course, and if the physical sciences are taught poorly in secondary school, then the chances any students will be interested in engineering diminish. Despite differences in methodology, the lines between engineering and the basic physical sciences are permeable borders, and students can move easily from one to the other. Calculations such as those for acceleration, concentrations, dispersions, stoichiometric ratios, saturation, and so forth are common to both. Engineering applications may employ these calculations to test the efficiency of various devices (and with typical attention to conservation), for example, while the basic sciences will use them more in the course of testing hypotheses about properties and actions of physical phenomena. Concepts such as mass spectroscopy, atomic adsorption, and ion chromatography will be illustrated in both chemical engineering and analytic chemistry labs. The point is that if one makes the investment in learning in the basic sciences as part of engineering education, then decides that the engineering is not as attractive, it is much easier to switch majors to a physical science than to the life sciences, social sciences, business, humanities, or arts.

The Social World of Engineering

But it is precisely at the point when a student finally decides to change fields, that the fundamental differences between engineering and scientific problems and problem-solving come into play. Science seeks to uncover the laws of nature, expand and deepen our knowledge of basic physical and biological phenomena, and create new materials or agents, thus establishing the foundations for applications. The constraints of clients, time, or resources are sometimes incidental, and often secondary, to these activities, depending on the organization in which they are carried out. Relying on the discoveries and creations of science, engineering seeks to delineate answers to client-posed problems within the constraints of time and resources, and then to select the optimum response from the range of possibilities. The client is the wild card, introducing ambiguity, idiosyncrasy, bias, and outright whimsiness. Academic science does not allow for such wild things, though research in most corporate and government laboratories is driven by sponsor interests—and even in such cases, "interests" are not "specifications." The best engineering solutions fit the culture in which they are applied: they are highly relativistic. A scientific response, on the other hand, pays attention to a nature that transcends culture.

Practicing engineers inhabit both an "object world" and a "social world" (Bucciarelli and Kuhn, 1997). Work in the "object world" is about relations and cases of behavior and properties of the physical phenomena that are the objects of design—buildings, machines, manufacturing plants, computer operating systems. The goal of object world work is to perceive and apply theoretical constructs to resolve empirical ambiguities and tensions into stable and replicable designs. The actual physical object, as Bucciarelli and Kuhn point out, is not necessary for object world work, even though the "craft aspects of work" dominate the values inherent in the design activities of engineers (Robinson and McIlwee, 1991, p. 405). But design problems exist because there is a "customer," even another division of the same firm. With the customer comes a social world, and processes of continual refinement through communication and negotiation as a project evolves. With both object and customer also come conditions of law, cost, organizational context, and other landscapes that must be defined, understood and traversed. The social world of engineering practice thus requires a very different kind of learning than that embedded in the application of mathematical models or in the 15 concepts that Warren (1989) pointed out were necessary to solve a paradigmatic problem of compressible flow in mechanical engineering.

We would do well to think further about clients and resources. High school students do not know what either concept is about; and neither do most college freshmen. But it is precisely because engineering tasks are carried out for clients that the issue of resources—their availability and cost—is part of project design. There are thus many answers to an engineering problem, but usually only one for a scientific problem. In this regard, too, engineering problems are solved within cultural constraints far more than are scientific problems. If a given culture believes that the gods of water are never to be disturbed, then the footings for bridges must be built on the shores, not on pilings sunk in the middle of rivers. Engineers will be hired to build such bridges, and the cultural assumptions become part of the problem the client poses and that the engineer must solve. The gods of water are a design criterion.

The dynamics of practice in engineering value design above all other activities (Perrucci, 1970; Noble, 1977; Whalley, 1986). The design process in any field of engineering reveals far more than analysis of the final product. Negotiation—not only with clients but among parties to development, crafting, and construction—is constitutive to the design process, and involves constant evaluation of trade-offs, including social and economic trade-offs. What are the virtues and limitations of approach X? shape Y? site D? Materials M and Q versus N and R? For both engineering and science, history is a constant guide; it is an intersection. Both scientists and engineers must ask, Who has tried anything like this before? Where? Under what circumstances? What happened? Why? But the answer the engineer gets reads more like a story with characters including accountants, legislators and sales managers than anything one reads in the sagas of scientific investigation.

I stress this distinction because engineering students have not been recruited by exposing them to the ways in which culture and personality enter design criteria for real world solutions to engineering problems. I think engineering would be attractive to more women, in particular, if the richness of practice, with all its contextual relativism, were the framework for education. To be sure, not all branches of engineering encounter design problems in which culture plays a significant role, but all branches encounter problems brimming with ambiguities and conditional situations that are analogous to cultural contexts. Well-crafted engineering design courses set with these assumptions can also assist students' psychological development (Pavelich and Moore, 1996).

Take, for example, some small-scale, low-tech projects currently used in the freshman design course by small teams (2-3 students) at the University of Wisconsin school of engineering:

Client	Project Outcome
Hospital	Adjustable stairs (run and rise) for patients who are rehabilitating
University Theater	New system to raise and lower backdrops
Dept. of Kinesiology	Handlebar grip for recumbent bicycle for use by physical therapy patients
Ski Resort	Mountain bike rack for different kinds of chair lifts
Agricultural Extension	Projects to reduce "stoop labor," for example, a Remay (cover cloth) furling and unfurling system

These projects were selected because (among other criteria) they "have at least 3 reasonable (to the faculty) solutions, and no obvious off-the-shelf commercial solutions . . . customer needs can be assessed . . .design criteria can be generated from customer needs . . . multiple design priorities exist and can be assessed." (Prof. Patrick Farrell, Department of Mechanical Engineering, University of Wisconsin, personal communication). There is nothing fancy here; but the projects are microcosms of the principles of engineering practice. They require learning context and culture of the client, negotiation, alternative designs, optimization, construction ("tinkering"), evaluation. This is not science.

These too brief reflections on engineering and science are intended to preclude some dead-end reasoning in the research on field attrition in engineering, particularly the reasoning that assumes a monolithic "S.M.E. culture" (Seymour and Hewitt, 1997). For a notable example, Astin (1993) hypothesizes that students who leave engineering for the physical sciences, particularly in schools where a comparatively large proportion (16+%) of students major in physical sciences, do so for the "status" or "prestige" of the pure field as opposed to the applied field (p. 256). The speculation is surprising in light of Astin's historical contention that people who start out majoring in engineering are motivated, in part, by the prospect of future financial rewards (something that cannot be said about physical science students). It is even more surprising in light of the depressed labor market for physical science graduates that emerged the mid and late 1980s, the period in which the subjects of Astin's *What Matters in College?* (as well as those in the High School & Beyond/So Cohort) were in school. Among other aspects of growth we expect in college are the discovery and confirmation of learning style and comfort with knowledge paradigms. It is more likely, I think, that some students who begin in engineering discover a greater degree of comfort with the experimental life of science. When they find out what engineering is, they may not like it. They may say, "if this is what engineers do, then it's not for me."

But culture and style may have a great deal to do with it, and national surveys, including the CIRP and the HS&B/So, do not capture culture and style. In Becher's (1989) analysis, students learn the rules of disciplinary culture fairly quickly, though not until graduate study in a discipline is that culture internalized. There are distinct communication patterns, idioms, symbolisms, sagas, behavioral norms. In undergraduate life, if students are uncomfortable with these "tribal" behaviors, they may go somewhere else. Engineers think heuristically, and talk about "quick and dirty" solutions. They sometimes worry about their academic status vis-a-vis theoretical science, and, in the engineering workplace, these status concerns translate into the benchmark of project management (Kunda, 1992; Robinson and McIlwee, 1991; Whalley, 1986). Students who prefer "elegant" solutions, bounded knowledge, and position in the academic hierarchy do not last long in the engineering labs. It's a matter of style, though one doubts many undergraduates could articulate it, and the student development literature tends to reduce style to Rotter's locus-of-control scales and similar measurements that lack the flavor of disciplinary culture. Deans and academic advisers who seek three-dimensional information to guide students might pay more attention to these cultural dimensions of the disciplines.

Part 4—Antecedents of the Engineering Path: Beginning the Story of Women and Men

So far, this exploration has delineated the nature of engineering paths and the various destinations arrived at by undergraduates, provided some detail as to what they studied on the way to these various destinations, and posted some road signs about the comparative textures of engineering and science paths. But undergraduate careers cannot be separated from their antecedents. When one looks backwards from engineering path history into the secondary school experience of students, the differential story of women and men begins to emerge. This story is not a simple one; there are many cross-currents, foremost of which is that of women's pre-college academic preparation for SMET majors and their actual choices of fields in college. As we will see, the women and men who reach at least the threshold of the engineering path look remarkably alike in terms of high school academic backgrounds, and very different from the women and men who did not attempt to reach toward an engineering program. If this observation sounds like common sense, it is.

Historically, the literature has approached gender issues in SMET fields with the deterministic paradigm of science itself. The preponderance of the research seeks predictive certainty and causal explanations. "The reason women do X is because they did/did not do Y or experienced/did not experience Z, etc." It is as if we can micromanage outcomes by identifying all observations of current and past conditions, reducing them to dichotomous variables, and making appropriate adjustments according to near-mathematical models. But we should have learned what science itself has learned in this century: that even within deterministic models there is much random behavior. This the essence of chaos theory. Chaos models are not linear, rather, reveal patterns, geometric-type relationships, diffusion. Minor changes in initial conditions, e.g. temperature or altitude, can produce vast oscillations in reactions. Local dynamics supersede global dynamics (Gleick, 1987), and we often do not understand how this happens.

The Math Path

How does the whiff of chaos theory enter into the account of engineering path models? Let us take one of the strongest lines in the research literature, one that I call "Math Path." It is commonly acknowledged that one's secondary school mathematics background is a powerful filter on participation in engineering programs. It is very difficult—though not impossible—to move from a high school "high" of Algebra 2 to a bachelor's degree in engineering, and yet, as table 14 demonstrates, about 21 percent of the women and 37 percent of the men who started on the engineering path at all started from that level—or less. In fact, among the completers on the engineering path, 33 percent of the men and 29 percent of the women (the difference is not statistically significant) emerged from high school with no more than Algebra 2. How did they do it? It turns out that half of them earned their degrees in engineering technology, for which, in the 1980s, as the transcript evidence suggests, one could present pre-calculus and statistics, or calculus at a later point in their program than is the case for engineering (Wolf, 1987). The others were obviously determined, and determination is one of those human characteristics that overcomes linear models.

Overall, the correlation between the highest level of mathematics studied in high school and bachelor's degree completion for the HS&B/So cohort was .525 (p < .0001), a very strong number for a relationship in which the outcome (the dependent variable) is an event far removed in time from the independent behavior, one that is slightly stronger for men (.537) than it is for women (.509)[12], and stronger for students in the lowest SES quintile (.420) than for students in the highest SES quintile (.368). But as soon as one contracts the universe to students who reach the threshold of engineering—as we have defined that threshold—the correlation weakens to .410, and, when one excludes students in 2-year engineering tech programs and those who are still enrolled in engineering, the correlation drops precipitously to .170.

In other words, once at the threshold of the engineering paths, one has already fulfilled a key condition in mathematics achievement, and the effects of prior study of mathematics diminish. Women arrive at the threshold with a slightly stronger (though, as table 15 demonstrates, not significantly different) mathematics profile than men, but it doesn't make much of a difference after that point—at least in terms of completing a bachelor's degree, whether in engineering or another field.

Heckel (1996), speculates that "for each discipline, enrollment is composed of a group of students which will consider only that discipline for academic study plus a fraction of an additional group that will consider a range of disciplines. The latter group accounts for the maxima and minima in degrees and enrollments as their academic interests fluctuate with time." (p. 16) If so, the critical question for any discipline in planning faculty or facilities is the size of the "fraction," or swing group. Engineering is one of those disciplines in which the confidence level in setting boundaries on that fraction is high. In fact, for all fields requiring competence in college-level calculus, if we know the proportion of freshmen who have successfully completed trigonometry in high school, if not pre-calculus, we can set a very strong boundary. Using a logistic regression in which the outcome is earning more than 4 credits in calculus in college, moving from a "high" of Algebra 2 in high school to trigonometry--a minor change in initial conditions--increases the odds by a factor of 2.3:1[13].

Table 15.—Highest mathematics studied in high school by all men and women who reached at least the threshold of the engineering path

	Men (85.4%)	Women (14.6%)
Calculus	24.0% (2.75)	29.0% (5.70)
Pre-Calculus	16.4 (2.20)	18.7 (5.10)
Trigonometry	22.6 (2.70)	30.6 (8.93)
Algebra 2	27.0* (3.14)	14.8* (4.57)
<Algebra 2	10.0 (2.09)	6.9 (3.94)

NOTES: (1) *$p \leq .05$ (2) Universe: all students who reached at least the threshold of the engineering path in 4-year college attendance patterns. Weighted N=139.5k
SOURCE: National Center for Education Statistics: High School & Beyond/ Sophomores.

Heckel's is a more credible approach to choice of major than those attempting to model causes of enrollment in a combination of expectations and values (see, e.g. Lips and Temple, 1990). To proclaim "causes of enrollment" implies that we can control choice. It is more likely that we can provide secondary school students with enough of a broad curricular momentum so that their choices are not constrained by lack of opportunity to learn. What we can also do is to protect equity in opportunity and socialization. Opportunity means what it says: at a moment in time, a generational moment, what were the real and (more importantly) perceived opportunities in education, career choice, labor market, etc.? Socialization would describe the contemporary cultural assumptions that children and adolescents acquire concerning roles and opportunities. The former can change while the latter lags, a divergence insufficiently recognized in multivariate analyses of choice of major (e.g. Wagenaar, 1984).

Science Background and Curricular Momentum

If the high school mathematics backgrounds of men and women who reach the threshold of the engineering paths in four-year colleges are remarkably similar, so are their high school science backgrounds. Compared to other students who attended 4-year colleges but never reached the threshold of engineering, however, they are very different. Drawing again on the high school transcripts of the HS&B/So cohort, table 16 focuses on core laboratory science courses in biology, chemistry and physics, and table 17 isolates chemistry, as, among the basic physical sciences, it was far more accessible than physics in U.S. high schools in 1982 (West, Diodato, and Sandberg, 1984)[14]. Both tables illustrate the similarities in science course-taking of men and women who reached the threshold and the divergence between their course-taking profile and that of women and men who made no attempt to study engineering.

The within-engineering threshold differences between men and women in both tables are insignificant, both statistically and substantively. Roughly 60 percent of both groups took three or more years of core laboratory science in high school; a smaller but roughly equivalent percentage of men and women took a second year of chemistry, which we can use as a proxy for Advanced Placement[15]. Among those who did not attempt an engineering curriculum in college, the differences in high school course taking, by gender, are slight, and statistically significant only for those who took two or three years of core high school science, with men having the edge. As Hoffer and Moore (1996) confirm for a more recent longitudinal studies cohort, gender differences are small not only in terms of the rates and types of mathematics and science course enrollments in high school (see also Madigan, 1997), but also in terms of teacher reports of instructional practices. Differences by SES, on the other hand, are more pronounced, though not overwhelming until one gets to the bottom quintile (in the HS&B/So data, 33.5 percent of those in the highest SES quintile took three or more years of core science in high school versus 27.4 percent of those in the 2nd, 3rd, and 4th quintiles; $t=2.85$; and 14.7 percent of the lowest quintile; $t=6.29$).

I stress secondary school curriculum in this analysis because, more than any other identifiable factor, it provides momentum, and in SMET fields, curricular momentum is worth far more than a grade point average or test score. Among all HS&B/So students who continued on to college, those with mediocre high school grades whose highest math was pre-calculus, for example, beat "A" students whose highest math was Algebra 2 to bachelor's degrees by 63 percent to 53 percent ($t=1.96$). Multivariate analyses reveal the same power of curricular momentum over grades. Controlling for SES, for all students who earned more than ten credits in college (i.e. the incidental students are out-of-scope for this analysis), the highest level of mathematics reached in high school accounts (adjusted R^2) for 23.1 percent of the variance in bachelor's degree completion rates. When one adds high school academic grade point average to the formula, the adjusted R^2 modestly increases to .273. If one limits the universe to those who earned more than 10 credits from 4-year colleges, the effects

Table 16.—Secondary school science curriculum of those who reached the threshold of the engineering path compared with those who did not reach the threshold

Years of High School Coursework in Core Science Fields:	Engineering Threshold		No Threshold	
	Men	Women	Men	Women
One/less	15.3%	12.1%	36.9%	36.5%
	(2.58)	(4.50)	(1.70)	(1.42)
Two	23.4	28.4	27.2	34.2
	(2.49)	(5.90)	(1.45)	(1.21)
Three	41.4	37.6	25.3	20.3
	(3.38)	(6.82)	(1.50)	(1.11)
>Three	19.9	21.9	10.6	9.0
	(2.69)	(6.19)	(1.12)	(0.88)

NOTES: (1) The universe consists of students who earned more than 10 credits from 4-year colleges, and for whom complete high school transcripts were accessible. Weighted N=1.2M. (2) Threshold students are 9.6 percent of the universe, and among them, 85.4 percent are men. Among "no threshold" students, 42 percent are men. (3) Standard errors of the estimates are in parentheses. SOURCE: National Center for Education Statistics: High School & Beyond/Sophomores.

decline (that is inevitable), but the relative impact of mathematics curriculum versus curriculum plus overall academic performance shows roughly the same ratio (the adjusted R^2s are .11 and .14, respectively).

The momentum hypothesis is in harmony with research showing that the minor differences in learned cognitive abilities between women and men are responsive to education and training. For example, Linn and Hyde (1989) demonstrated that the largest gender differences in mathematics are observed in solving word problems, and that, despite women's edge in verbal skills, the variance reflected disparities in high school course participation rates in physics and chemistry. While the data in tables 16 and 17 do not evidence these disparities,

Lynn and Hyde's conclusions have the ring of common sense. Participation, exposure, involvement—these lead to improved performance, and there is no question that in the decade following the passage of Title IX of the Education Amendments of 1972 more young women were both exposed to and involved in both science and higher levels of mathematics, and that this participation is reflected in the profiles of the high school class of 1982 (the HS&B/So).

Table 17.—Secondary school course taking in chemistry of those who reached the threshold of the engineering path compared with those who did not reach the threshold

	Engineering Threshold		No Threshold	
	Men	Women	Men	Women
Years of High School Coursework in Chemistry:				
Less than one	23.1%	16.9%	46.4%	50.0%
	(2.97)	(5.18)	(1.74)	(1.53)
One	62.0	65.2	48.0	44.4
	(3.38)	(7.27)	(1.70)	(1.53)
Two	14.8	17.9	5.6	5.6
	(2.37)	(4.98)	(0.83)	(0.73)

NOTES: (1) The universe consists of students who earned more than 10 credits from 4-year colleges, and for whom complete high school transcripts were accessible. Weighted N=1.2M.
(2) Standard errors of the estimates are in parentheses.
SOURCE: National Center for Education Statistics: High School & Beyond/Sophomores.

Academic Performance

As for overall academic performance in high school, women have a clear advantage (as they do in higher education). To measure performance in such a way as to minimize school effects, I created a variable that combines grade point average in academic courses only with high school class rank (the correlation between the two is .81) in quintiles. Table 18 sets forth the proportion of students in the highest two quintiles by engineering path destination. While all engineering path students in 4-year college attendance patterns were stronger students in secondary school than their peers, the migrants begin to emerge as a relatively weaker group—in terms of secondary school background—in the engineering path universe.

46

The difference between "migrant men" and "threshold men" in table 18 are not statistically significant, but those between "migrant women" and women in the other engineering path groups are significant, and the same pattern can be observed in test scores.

Table 18.—High school academic performance of HS&B/So students by ultimate engineering path status

	Percentage of All College-Going 1982 High School Graduates in Highest Two Quintiles of Performance				
	All	Men	Women	‖	% of ALL
ALL	57.8	49.9	64.6	‖	**100.0**
No engineering path	56.6 (.089)	46.1 (1.34)	64.3 (1.13)	‖	**91.1**
Threshold only	74.9 (6.35)	70.7 (7.36)	100.0 (N.A.)	‖	**1.3**
Migrants	73.2 (5.59)	76.7 (5.87)	57.1 (8.13)	‖	**1.5**
Completers	85.6 (2.56)	84.8 (2.80)	93.0 (6.63)	‖	**4.0**
2-Year EnginTech	35.4 (5.38)	32.4 (5.33)	--	‖	**2.1**

NOTES: (1) The universe includes all students for whom high school academic performance could be computed and who earned more than 10 college credits. Weighted N=1.8M. (2) Standard errors of the estimates are in parentheses. (3) -- N insufficient to produce a reliable estimate.
SOURCE: National Center for Education Statistics: High School & Beyond/ Sophomores.

There are two sets of test scores in the HS&B/So database. One consists of an enhanced "mini-SAT" (the test includes sections on science and civics in addition to reading, writing, vocabulary, and mathematics) that was administered to 91 percent of the sample in the 12th grade. While this is not a high stakes test, the results for the general population are highly correlated with the SAT (Heyns and Hilton, 1982). The other consists of a composite SAT score for 57 percent[16] of the sample who took either the SAT or ACT at any time through grade 12 (ACT scores were converted to the SAT scale). Our data on SATs show the following for the tripartite destinations on the engineering path:

Table 19.—SAT/ACT scores of engineering path students

	Male	S.D.	s.e.	Female	S.D.	s.e.
Threshold only	1016	150	1.90	1093	172	4.74
Migrants	1111	187	1.86	938	227	5.42
Completers	1092	191	1.22	1112	150	3.02

NOTES: (1) Universe: All students who reached at least the threshold of the engineering path, excluding those in 2-year engineering tech programs, and for whom either SAT or ACT scores were available[17]. Weighted N=95k. (2) Standard errors of the estimates have been adjusted for design effects.
SOURCE: National Center for Education Statistics: High School & Beyond/ Sophomores.

These comparisons, all of which are significant at the $p \leq .05$ level, present a more complex picture than those presented by the National Science Foundation for "engineering students" of the same period (NSF, 1986) or the SAT mean of 1103 offered by Kroc, Howard, Hull and Woodard (1997) for all entering engineering students in a sample of public research universities of a slightly later period (1988 and 1990). The HS&B/So women who completed engineering degrees had slightly higher SATs than male completers and were more uniform in performance (as evidenced by a smaller standard deviation), an observation confirmed by a Purdue University study during the same period that the HS&B/So was in its undergraduate phase (Epstein, 1991). But women who left engineering performed much worse than men on the SAT and, as a group, evidenced greater variance in performance, an observation strongly confirmed by the results of the enhanced "mini-SAT"[18]. As McIlwee and Robinson (1992) observe, for women more than men in a field with the gender imbalance of engineering, "only the academically strongest are likely to survive" (p. 75).

High school curriculum, academic performance, and test scores—all traditional input measures—tell us that students who arrive at the threshold of the engineering path have a more powerful curricular momentum than their peers and a performance profile indicating that they are more likely to complete bachelor's degrees in any field. The women who arrived at the threshold brought to the table of higher education the same strong academic profiles and curricular momentum as did the men. Remember, though, that those who reach the threshold are an elite group: they constitute only four percent of the high school graduating class and nine percent of those who continued on to college.

Student Choice as Wild Card

Let us turn the dependent variable on its head as a prelude to investigating the actual choice of an engineering major. That is, let us find all the students in the HS&B/So high school graduating class of 1982 whose records exhibit the same curricular momentum and high quality academic profile as did those who reached at least the threshold of the engineering path, but who did *not* choose the path. When they were in grade 12, what major did they indicate they intended to pursue? As they entered higher education, in what ways did they behave differently from those who reached the engineering path? Table 20 presents this high talent group (they also appear under a similar rubric in table 26, but are included with all non-engineering path students in that presentation), and we note, as background, that 81 percent of them ultimately earned bachelor's degrees, nearly double the rate for all students who earned more than 10 credits in higher education (Smith, *et al*, 1996, p. 25).

Note that nearly 12 percent said they were going to major in engineering or architecture but never made the effort; half of them did not enter doctoral degree-granting institutions, hence were less likely to begin college in a school with an engineering program. A quarter of these students eventually earned bachelor's degrees in computer science or mathematics, but no other majors except economics claimed more than 10 percent of the group. This is a very diffuse pattern of behavior.

It is just as diffuse for the 20 percent who said that they were either "undecided" or "pre-professional." The two categories can be lumped together because pre-medical students or pre-law students who say their college major will be "pre-professional" have not really made up their minds. Indeed, there is no distinct pattern in the fields in which they ultimately earned their bachelor's degrees (business, physical sciences, economics, life sciences, and humanities accounted for half the degrees they earned, and in roughly equal proportions). As Kroc, Howard, Hull and Woodard (1997) demonstrated with samples of entering freshmen in 1988 and 1990, the "undecideds" who earned degrees within five years also scattered across the disciplines, though business and the social sciences took half of them.

Women constituted a disproportionately high share (59.7 percent) of this high talent group with curricular momentum in science and mathematics. In weighted numbers, that's 78,000 female students from a single high school graduating class who continued on to college, but they were dispersed over 1,000 institutions, most of which do not offer engineering programs. The preferences of these potential SMET recruits were concentrated in two areas: undecided/preprofessional and health sciences/services. The undecideds/pre-professionals, as we have observed, scattered after college entrance, and women's behavior was not different

Table 20.—Anticipated field of postsecondary study of 1982 high school graduates with high quality curricular and performance profiles but who were not on the engineering path in college

	Men	Women	All
Pre-Professional/undecided	18.3% (3.89)	21.7% (4.00)	20.3% (2.95)
Health sciences/services	3.7 (1.96)	21.5 (4.34)	14.3 (3.02)
Engineering/architecture	16.1 (4.10)	8.8 (2.46)	11.7 (2.07)
Computer sci/mathematics	14.9 (4.11)	6.9 (2.23)	10.1 (2.14)
Business	9.8 (2.39)	10.2 (2.69)	10.0 (1.93)
Life sciences	12.5 (3.28)	8.7 (2.40)	10.2 (1.92)
Physical sciences	10.8 (3.30)	6.3 (2.06)	8.2 (1.85)
Other	13.9 (3.79)	15.9 (2.97)	15.2 (2.39)
Percent of ALL	**40.3**	**59.7**	**100.0**

NOTES: (1) Universe: all students who graduated from high school on time, completed mathematics courses at the level of trigonometry or higher, took at least three years of core laboratory science including at least one year of chemistry, were in the highest two quintiles of academic GPA/class rank, and the highest quartile of the enhanced "mini-SAT" given to all seniors, continued their education after high school, earned more than 10 credits, but never reached the threshold of the engineering path. Weighted N=130k. (2) Standard errors of the estimates are in parentheses.
SOURCE: National Center for Education Statistics: High School & Beyond/Sophomores.

from men's in this regard. But those who indicated a preference for college programs in health sciences/services (nursing, physical therapy, nutrition, public health, health administration) persisted to a high degree on those paths: over half the women who indicated this preference in the 12th grade earned bachelor's degrees in these fields and another third earned associate's degrees, principally in nursing. As we will note in Part 5, these fields have a high "staying power" in students' occupational vision (see table 21). Most of these fields also exhibit gender imbalances, though on the female side. As we are about to see, in the finite-glass game of competition among academic fields for highly talented students, curricular momentum does not have inevitable destinations; it is confounded by student choice, and student choice is often a wild card.

Part 5—Choosing the Engineering Path

The literature on choice in engineering is difficult to sort out for three reasons. The first involves a conceptual substitution: "career" for "academic major." The second is temporal: the selection of different moments of choice (pre-college, on entrance to college, early in a college career). The third is the tendency of analysts to aggregate engineering with other fields. Each has significant effects on the universe of people under study. And each has significant effects on the analysis of field attrition among men and women.

Career versus Major

Using a national cohort from the Cooperative Institutional Research Project (CIRP) that entered four-year colleges in 1985 and was followed up in 1989, Astin (1993) indicated a moderately high correlation (.49) between freshman and follow-up choice of engineering as a *career*, but also notes that of entering freshmen planning to become engineers, only 36 percent held on to those plans. At the same time, the field retention rate for engineering *majors* over the same period was 57 percent (Astin and Astin, 1993). Quite simply put, these data illustrate the difference between "career" and "major," and that field attrition is not equivalent to career plan changes (for a similar observation, see Sax, 1994).

In his baseline work, *Four Critical Years* (1977), Astin developed an analysis of stability and change in career choice during the college years using ten general occupational fields accounting for the preferences of approximately two-thirds of entering freshmen. He looked forward in time from freshman to senior year, and found the largest decline among those who said they were going to be engineers; but when he looked backward from the senior year and asked what proportion of students with specific career goals held the same goals as freshman, engineering ranked first (74 percent), followed by school teaching (64 percent), nursing (62 percent) and medicine (51 percent). In other words, these occupational field goals attract a core of firm believers and shed those whose initial commitments may have been more casual and/or those whose abilities did not match the academic demands of preparation for the occupation at issue. Astin did not put it this way, but engineering as an occupation, unlike medicine or college teaching, translates directly into engineering as an undergraduate major. In order to keep a clear distinction between major and choice of career in science and technological fields, Astin and Astin (1993) developed a more convincing tripartite division of career choice: engineer, research scientist (which includes college teacher), and scientist-practitioner (which includes physician).

In 1980, 1982, 1984, and 1986, the HS&B/So surveys asked students to indicate their anticipated occupation at age 30. Because this longitudinal study began in grade 10 and covers students who do not go to college and those who attend sub-baccalaureate institutions, the range of occupational choices is broader and more generalized than that of the CIRP surveys. The categories of response are 16 aggregates used by the Bureau of Labor Statistics and the Bureau of the Census and unfortunately do not allow us to identify engineering separately (engineering is buried in a category called "miscellaneous professions"). But

Table 21.—Consistency of vision: anticipation occupation versus actual occupation at age 27/28, in selected occupational clusters

1991 Occupation	Percent with Consistent Vision 1982, 1984, 1986	Percent with Constantly Changing Vision 1982, 1984, 1986
Medical professionals (MDs, RNs, etc.)	43.2 (3.62)	15.6 (2.65)
Engineers, architects, computer programmers	41.0 (3.92)	10.0 (2.26)
Managers: manufacturing, agriculture, construction	32.5 (4.24)	22.2 (3.88)
Educators: school, college, other	31.6 (3.27)	19.8 (2.94)
Legal and financial professionals	31.2 (2.01)	22.4 (3.10)
Communication, arts, entertainment	30.7 (4.13)	16.5 (3.62)
Other technical	20.5 (2.94)	28.8 (3.45)
Clerical	18.5 (1.65)	33.9 (2.11)
Managers: retail, other	18.2 (1.79)	33.4 (2.09)
Military and protective service	16.7 (2.73)	29.2 (3.61)
Craftsmen, mechanics, and skilled operatives	16.7 (1.58)	35.9 (2.00)
Marketing and sales	13.2 (1.68)	30.8 (2.36)

NOTES: (1) Universe: all High School & Beyond/Sophomores who answered questions about anticipated occupation at age 30 in 1982, 1984, and 1986, for whom occupational status in 1991 could be determined, and who earned more than $5,000 in 1991[19]. Weighted N=2.39M. (2) Standard errors of the estimates are in parentheses.
SOURCE: National Center for Education Statistics: High School & Beyond/Sophomores.

because the HS&B/So includes full long-term labor market histories (the CIRP longitudinal studies do not) what we can do to illustrate both the difference between firm believers and casual participants, and to underscore the potential of students' degrees of commitment as influencing persistence in science and engineering (Grandy, 1995), is to construct a corollary to commitment in "consistency of occupational goal." This latent variable emerges in the HS&B/So data only in light of actual occupation in 1991 (at age 27/28). If a student gave the same response to the 1982, 1984 and 1986 questions about anticipated occupation, for example, "professional practice," then the student was exhibiting at least a general consistency of occupational vision. Table 21 sets forth the proportion of HS&B/So participants working in 12 occupational clusters in 1991 who exhibit this consistency of vision, and, for the same occupational clusters, the proportion who were least consistent in their occupational goals (that is, they gave three different answers to the same question in 1982, 1984, and 1986).

Table 21 clearly shows that engineering is in an occupational cluster attracting a high proportion of people who had a consistent occupational goal and a low proportion of people who were constantly changing their career objectives. Along with architecture and computer-related professional work, engineering had a high degree of staying power for the HS&B/So cohort. The same can be said for the medical professional cluster.

High school students do not exercise rigid rational choice decision models in their occupational aspirations (Wagenaar, 1984). Those students who are not confused about what it means to be an engineer, and who are committed to professional life in applied science are going to be very consistent in their vision, and this consistency, we can reasonably hypothesize, will drive their academic life. Even those who may have been confused when they were in high school about the nature of engineering but, on the occasion of clarification (or epiphany) in a freshman design course, decided that they liked it, will adhere with that vision. And if you believe that the outcomes of completing a degree in engineering are positive, you will reinforce your own efforts, hence be more likely to persist to a degree (Hackett, Betz, Casas, and Rocha-Singh, 1992). On the other hand, for example, the vast and amorphous area of our economy I call "buy/sell" (in the occupational taxonomy used above is labelled "Marketing and Sales") attracts people who, quite frankly, cannot make up their minds what to do in worklife. It is the labor force "default," so to speak.

Bucks and Salary

Two decades ago, Polachek defended the hypothesis that the personal value placed on monetary rewards (pecuniary values) were the most powerful determinant of field choice (Polachek, 1978). Following in this tradition, Heckel (1996) argued that students make decisions, in considerable part, on the basis of perceived financial advantage in an occupation, and that their perception is based on lagging experience, i.e. what happened in the labor market in the previous four or five years. He also argues, albeit lightly, that engineering educators should encourage persistence through exposure to the business aspects of engineering and a mentoring that reminds students of the economic climate.

Other noted literature (Astin, 1993; Seymour and Hewitt, 1997; Astin and Astin, 1993) also cites a corollary of the pecuniary determinates of career choice in the "materialism" of engineering students. In light of this persistent theme, it is fortuitous that the HS&B/So database provides the opportunity to test the case, though in a different manner than did Wagenaar (1984)[20]. On four occasions between 1980 and 1986, the surveys asked an identical series of "life-value" and "work-value" questions. From these, one can develop indicators that set the importance an individual ascribes to financial well-being, making a lot of money, and salary in an occupation, against the strength of the individual's commitment to community, family, learning on the job, and so forth. The ratios that emerge are of the importance of money: in the context of other life-values (the "BUCKS" ratio) and work-values (the "SALARY" ratio).

Table 22.—Relative importance of money in work and life for HS&B/So students, by selected engineering path status, as seniors in high school and 4 years later

| | Against Work Values | | Against Life Values | |
	1982	1986	1982	1986
No engineering path				
men	.911 (.245)	.948 (.242)	1.01 (.288)	.936 (.294)
women	.879 (.233)	.929 (.225)	.926 (.289)	.890 (.276)
Migrants				
men	1.02 (.217)	.944 (.214)	1.06 (.356)	.987 (.308)
women	.826 (.356)	.979 (.211)	.887 (.238)	.901 (.217)
Completers: grad school				
men	.886 (.194)	.920 (.226)	.953 (.255)	.856 (.258)
women	.950 (.227)	.906 (.289)	.911 (.379)	.836 (.365)

NOTES: The universe consists of all students who earned more than 10 credits in higher education and who answered all questions about values in work and life in both 1982 and 1986. Standard deviations are in parentheses (for comment on standard errors of the estimates, see Technical Appendix).
SOURCE: National Center for Education Statistics: High School & Beyond/ Sophomores.

Table 22 takes both variables at two moments in time for the High School & Beyond/So cohort: as seniors in high school (1982) and four years later (1986), and illustrates the point with a selection of populations. In table 22, a ratio of 1.0 is a balanced position: the respondents valued money equally with other goals of life (for example, giving one's children better opportunities, having close friends, being a leader in one's community) or work (for example, opportunity to use education on the job, autonomy in work, job stability). A ratio less than 1.0 indicates a lesser weight for money; a ratio greater than 1.0 is the converse. The range of the mean ratios is fairly narrow, so one pays more attention to the direction and magnitude of change. Most of the within-category (e.g. migrants/work values/1982) differences between men and women are statistically significant, as are most of the within-gender-category (e.g. migrants/men/life values) differences between the two time benchmarks[21]. Two groups stand out in terms of the diversity of change. Among "migrants," the standard deviations contracted in all categories between 1982 and 1986, indicating that, however diverse this group might be on other counts, its vision of the relative importance of money became more uniform in time. Among those who completed engineering degrees and continued on to graduate study, however, the weightings of the importance of money became more diverse between 1982 and 1986 (the standard deviations rose in three out of four possible pairs). Nothing is simple in an analysis that distinguishes destinations on the engineering path.

What we see in general of all men and women has been noted previously (Adelman, 1994): while men value money more prior to college entrance (1982), the "money values gap" closes in time. For both sexes, too, while the value of money gains in the context of worklife, it declines in the context of a more holistic vision of life that includes family, community, friends, and leisure time.

Table 23.—Mean earnings at age 27/28 (1991) for all HS&B/So students who continued their education after high school, in selected occupational clusters

Cluster	Mean Earnings in 1991	s.e.
Engineers, scientists	$36,408	$ 977
Doctors, lawyers	36,056	2,833
Business service professionals (e.g. accountants)	34,443	1,275
Computer-related	31,898	982
Licensed medical professionals (e.g. RNs)	30,263	1,008
Marketing and sales	29,153	1,007
Military/protective service	28,345	1,226
Crafts/skilled operatives	26,249	834
Communication/arts	24,096	1,367
School teachers	21,450	781

NOTE: Standard errors of the estimates have been adjusted for design effects.
SOURCE: National Center for Education Statistics: High School & Beyond/ Sophomores.

The evidence is thus mixed as to whether students choose initially to pursue an engineering degree because they perceive the personal economic rewards of engineering practice to be great. Heckel (1996) uses time-series data on starting salaries and volume of industrial interviews with graduating seniors as Michigan Technological University to argue that they do. If so, as table 23 confirms, the students are not wrong in their judgment of the labor market: the HS&B/So students who were practicing engineers at age 27/28 in 1991 were, in fact, in the highest salaried occupational cluster.

Choice of Major: When Do You Ask the Question?

One of the methodological issues that bedevils the study of academic choice and confounds enrollment planning is the timing of the question. Coupled with the timing of the question are the categories of response. Ask a high school senior what he/she plans for a college major and include in the possible responses such categories as "pre-professional" and "technical," and it becomes very difficult to predict who may concentrate in which fields two or three years later, when concentration patterns begin to emerge (see table 1). The High School & Beyond survey included such categories. When one examines the college records of what "pre-professional" students actually studied, there are no distinct disciplines that dominate the histories. The portrait is a scatter-plot. The students who checked the box for "pre-professional" possessed but vague ideas; they were, as previously observed, undecided. Ask the same question of entering college freshmen, as the CIRP surveys do, and the responses take on a higher degree of validity, though if one includes categories such as "pre-med" among the possible responses, one is not describing actual majors. As previously noted, with respect to major, the "pre-meds" are, in fact, undecided, too.

The contention of this monograph is that the pre-college or entering college question, while a valuable component of the generic radar screen watched by provosts and deans, is not as important to understanding the engineering path as are the actual course-taking behaviors of students. This contention might be tested in the same manner that Astin (1993) tested career choice, with the additional advantage, though, of transcript records that allow us to distinguish among four categories of students: those who never attempted an engineering curriculum, those who reached the threshold, the migrants, and the completers. What do we find if we match what they said they were going to major in when they were seniors in high school against their actual behavior. First, it is extremely important to note that 40 percent of those who eventually earned more than an academic year's worth of credits from a 4-year college and who said they intended to major in engineering or architecture never reached the threshold of the engineering path. 40 percent! They did not even try. They are a distant potential part of the field, but they cannot be judged as migrants because they never reached a position from which to migrate.

But looking forward, as Astin found, is not as profitable as looking backwards. Table 24 asks the following question: given the final destinations of 4-year college students in the HS&B/So cohort, what did they tell us they were going to major in when they were seniors in high school?

Table 24 reveals, first, that students who reached *any* stage of the engineering path overwhelmingly intended to major in engineering when they were in the 12th grade. Just as Astin found with the 1995–1989 CIRP sample, seven out of ten HS&B/So students who completed degrees in engineering intended to do so when they were seniors in high school. Second, we note that the alternative intentions of students who reached any destination on the

Table 24.—Probable major indicated in grade 12 by HS&B/So students who later enrolled in 4-year colleges, by engineering path status

Probable Major as Indicated in 1982	Engineering Path Status in 1993				
	No Engin Path	Thresh-hold Only	Migrants	Completers	**ALL**
Engineering or architecture	5.8% (0.38)	63.4% (7.73)	59.5% (6.49)	71.2% (3.53)	10.2% (0.51)
Computer science or mathematics	9.1 (0.57)	16.0 (5.76)	15.1 (4.56)	11.0 (2.54)	9.4 (0.55)
Life, health, or agricultural sci	14.5 (0.69)	-- --	4.2 (.262)	3.0 (1.10)	13.7 (0.64)
Physical science or other technical	6.4 (0.50)	12.4 (5.34)	5.2 (2.27)	7.1 (2.11)	6.5 (0.47)
Undecided	16.7 (0.70)	-- --	8.8 (3.77)	5.0 (1.95)	15.8 (0.66)
Other	47.6 (0.94)	-- --	7.3 (3.19)	2.7 (1.11)	44.5 (0.94)

NOTES: (1) Universe: all students who answered questions about planned college major in 1982, who subsequently earned more than 10 credits from 4-year colleges and who were not in the 2-year engineering tech program category or the "still enrolled in engineering" category at age 30. Weighted N=1.6M. (2) -- N is insufficient to produce a reliable estimate. (3) Standard errors of the estimates are in parentheses. (4) Columns may not add to 100.0% due to rounding or low-N cells.
SOURCE: National Center for Education Statistics: High School & Beyond/Sophomores.

engineering path were concentrated in computer science/mathematics and physical sciences/other technical fields. Only among the migrants did a sizeable proportion (20 percent) fall in other categories (including "undecided"). Lastly, the proportion of this population that is listed as "undecided" in the 12th grade (15.8 percent) is higher than indicated in most other studies, and, I would argue, is a more accurate reflection of reality (in table 20, isolating the best prepared students, the portion who are *de facto* "undecideds" is even higher, 20.6 percent).

Women, Men and the Choice of Engineering

Student perceptions of occupations involve a good deal of sex stereotyping, even though the degree of stereotyping declined from the 1970s to the late 1980s (White, Kruczek and Brown, and White, 1989), and even if some of the stereotyping does not agree with actual degrees conferred (e.g. veterinarians and pharmacists were clearly on the "masculine" side of the scale used by White, Kruczek, Brown, and White even though women have received the clear majority of degrees in both fields for a decade or more[22]). Using Holland's (1985) classic classifications, engineers and architects have been persistently perceived as belonging to "realistic" occupations, and "realistic" is the most "masculine" position. Physicists and chemists, by comparison, are classified as "investigative" occupations, and these, while on the "masculine" side of the scale, are closer to the gender-neutral position.

The literature on choice in scientific fields often draws on the influence of parents, teachers, and peers. There are many variations on these themes, but the most persistent are about parents. While Seymour and Hewitt's (1997) elite group of interviewees did not weight "family tradition" as a significant factor in choosing a SMET major, Astin and Astin (1993) demonstrated positive influences of "occupational inheritance" on the choice of careers in science, medicine, and engineering. Fitzpatrick and Silverman (1989) found that support of both parents was important for women choosing engineering, but not important for women choosing science, and that the influence of high school teachers on women choosing science (but not engineering) was stronger than that of parents.

The High School & Beyond categories for parental occupations are not very helpful in this matter (they do not include "engineer" or "scientist," for example). However, as an indirect proxy for parental support, we can use the HS&B surveys to match parental expectations for their children's ultimate level of educational attainment by student choice of major as indicated in grade 12. What we see of this match in table 25 is that women who intended to major in engineering or architecture enjoyed the highest degree of parental support for bachelor's degree attainment among all women—or men—who intended to major in any other field (except education and the social sciences, where the differences between men and women are not statistically significant). The data do not tell us directly whether parents are supporting the field choices of their children, but the indirect evidence of student secondary school attainment profiles suggests that, in some cases, table 25 reveals less than full enthusiasm. The rates of parental support for students who intended to major in the physical sciences, a group whose high school curricular and performance profiles were very strong and who, in fact, ultimately enjoyed a 76.2 percent bachelor's degree attainment rate, are particularly troubling—for both men and women.

When they were asked for their anticipated college major in the 12th grade, only 5.4 percent of the HS&B/So women—compared with 22.8 percent of the men—who first entered 4-year colleges indicated engineering or architecture. Once enrolled, however, only 2.8 of women entering 4-year colleges—compared with 17.6 percent of the men—reached the threshold of the engineering path. If the parental influence hypothesis is as strong as we are sometimes lead to believe, that shrinkage is superficially counter-intuitive. Something else, however, was happening. It is important to note that this transitional period group is not large (the weighted N was 80,324), but that no other national database can get at it.

Table 25.—Parental support for student degree attainment, by intended major field

	Percentage whose parents expected them to earn a bachelor's degree or higher		
Major Field Choice in Grade 12:	Women	Men	All
Life sciences	77.6 (6.81)	76.1 (5.15)	76.8 (4.14)
Engineering/architecture	96.1* (2.34)	87.0* (2.11)	88.4 (1.83)
Computer science/math	71.2* (4.49)	84.5* (2.79)	78.5 (2.49)
Physical sciences	61.1 (6.44)	54.8 (5.86)	57.0 (4.54)
Health sciences/services	68.4 (3.39)	79.7 (10.8)	69.7 (3.24)
Business	75.1* (2.60)	85.9* (2.28)	79.7 (1.78)
Education	92.3 (2.66)	92.0 (4.48)	92.2 (2.78)
Humanities and arts	78.0 (3.55)	79.7 (4.40)	78.5 (2.79)
Social sciences	93.1 (2.29)	94.0 (2.80)	93.4 (1.79)
Undecided/pre-professional	80.4 (2.44)	83.4 (2.83)	81.7 (1.87)
Other	71.2* (4.69)	86.2* (4.68)	75.6 (3.66)
All	**77.4 (1.15)**	**82.6 (1.16)**	**79.7 (0.85)**

NOTES: (1) * Comparisons of women and men are significant $p \leq .05$. (2) The universe consists of students who indicated an intended college major when in grade 12, who also indicated their parents' expectations for their education, and who subsequently entered higher education and earned more than 10 credits. Weighted N=1.32M
SOURCE: National Center for Education Statistics: High School & Beyond/Sophomores.

Those who abandoned their intentions to major in engineering between the 12th grade and the time for declarations of major were part of what Heckel (1996) called the "swing group." When we look at this group with the full power of high school and college records in table 26, we find considerable differences between men and women, all statistically significant.

The men who abandoned their intentions to major in engineering between the 12th grade and college were, on the whole, a relatively weak group, and were cut short by lack of curricular momentum. The women had sufficient curricular momentum and attainment profiles to at least explore engineering, but chose not to do so. For men, one can reasonably assume that college academic advisers will not encourage students with weak high school mathematics and/or science preparation to major in engineering, let alone when combined with mediocre academic performance. At this transitional stage, curriculum—and curricularly-related experiences—are playing a role in the continuous filtering of choice. For men, this conclusion is obvious. For women, it is not.

Table 26.—Contrast between men and women who abandoned their intention to major in engineering between the 12th grade and early college experience

High School Background	Men		Women	
Highest mathematics was more than Algebra 2	40%		66%	
Took 3 or more years of core laboratory science	30		51	
At least 1 AP course	22		37	
Ranked in top 40% of high school graduating class	48		86	
Scored in the top 25% of 12th grade "mini-SAT"	40		64	
Postsecondary Attendance				
No 4-year college	27		12	
Took calculus in college	24		53	
Grades of C- or less		42		23
Earned Bachelor's	39		62	
In SMET Fields		6		21

NOTES: (1) The universe consists of all students who indicated an intent to major in engineering or architecture when they were in the 12th grade, who never reached the threshold of the engineering path in higher education, but who earned more than 10 credits in college. Weighted N=80,324. (2) All comparisons of men and women in this table are significant at p. ≤ .05.
SOURCE: National Center for Education Statistics: High School & Beyond/Sophomores.

Multivariate analysis provides some support for the case. When the dependent variable is reaching the threshold of the engineering path and four independent variables—highest level of mathematics, completing three of more years of core science, ranking in the top 40 percent of one's high school class, and socioeconomic status—are regressed on it in a stepwise manner, the adjusted R^2s for the men who abandoned their intention to major in engineering are .051, .060, .062, and .063 while those for the women are .022, .026, .027, and .027 respectively. What this means is that the curriculum model is stronger for men than women and that neither grades nor SES add anything to the explanatory power of the model. The men were weaker students, so curriculum (particularly the highest level of mathematics studied in high school) made more of a difference for them. For women, something else is at work that no regression equation will uncover.

My working hypothesis is that though high school curriculum provides momentum toward a college discipline, the culture of the discipline may have more to do with undergraduate persistence and attrition than anything a student's parents ever said. This hypothesis requires us to return to some of those differences between engineering and science, with particular attention to the way they affect women's choices.

Women and the Culture of Engineering

One of the problems in the traditional literature is that the analysis of sex differentials and inequality in scientific careers is grounded in the sociology of academic science, not the practice of engineering. Who attends graduate school and works with which eminent professors (Lodahl and Gordon, 1973), who publishes early, who wins grant support, who publishes often and on what (Pfeffer, Leong, and Strehl, 1977), who wins prizes and other recognition (Fox, 1984)—these questions are the stuff of inquiries into the inequality of scientific production and productivity. But the organization, environment, rules and processes of academic science are not those of engineering, and these features of academic science seem rather tangential to the story.

There is a growing constructive alternative literature on the culture of engineering, particularly as it applies to women's experience, a literature that branched from a more generic line of inquiry into women's encounters with the culture of organizations as they moved into "non-traditional" jobs in the 1960s and 1970s (for a notable example, Kanter, 1977). This literature reveals many cross-currents and paradoxes that call out for unravelling. Bailyn (1987) for example, pointed out that even women with outstanding undergraduate careers in engineering expressed ambivalence toward their technical expertise in the workplace. Grandy's research (1995) shows that, across all SMET fields, the reasons women leave involve not academic achievement or college pathways but a lesser degree of commitment and ambition in science: "Whatever reasons female students may have for leaving . . . they are apparently not intellectual reasons but reasons based on what they want to do with their lives." (Grandy, 1995, p. 29), a conclusion thoroughly seconded by Tobias (1990) and Seymour and Hewitt (1997).

61

This neo-realistic argument concerning women's under-representation in fields such as engineering attributes the phenomenon not to poor preparation, rather to alienation and lack of interest. In some hands (e.g. Lips, 1993), this observation forms the grounds for a sophisticated analysis of the "moment" of divergence in major/career choice for men and women. By "moment" is meant a point of high "affirmation" of interest in SMET that affects men's actual participation but not that of women. For women, then, interest in mathematics and science does not necessarily translate into majors or careers, and this dissonance is more pronounced in engineering/physical sciences than in mathematics/ statistics/computer science (Lips, 1993). Given Seymour and Hewitt's (1997) conclusion that women who entered SMET fields were more influenced by others (parents, teachers, counselors) in their initial choice than were men, or chose a path to please these others, it is not surprising that they fall away more easily: the choice was not intrinsic, and the interest not deep enough to be affirmed.

McIlwee and Robinson (1992) provide an important analytical framework for our appreciation of women's experience as engineers by adding consideration of power, style, and conflict in the workplace. In this view, the management of relationships and the acquisition of organizational resources (both formal and informal) are "every bit as valuable as an academic degree or technical expertise" (p. 16), even though technical knowledge and creativity remains the touchstone of power in the production processes of engineering. The culture of engineering workplaces in which bureaucracy is weak, engineers gain power through aggressive behavior, and shop talk is dominated by accounts of tinkering with things is a culture that works against women. Why? The features of adolescent experience that turn up in the narratives recounted by Seymour and Hewitt (1997), McIlwee and Robinson (1992), and Robinson and McIlwee (1991) are very consistent on this count: women chose engineering as a path because they were "good in math and science," while men chose engineering because they played with machines and circuits, they deconstructed and rebuilt, they "tinkered". The power in workplaces with informal organization goes to the tinkerers—until they either burn out or move into management roles for which they are not prepared (Kunda, 1992).

This culture, McIlwee and Robinson are quick to add, is not necessarily ascendent in all engineering organizations. Women do better in those organizations where it is less pervasive, in organizations that are more bureaucratic and in which engineers themselves are less powerful. Sonnert (1996) would agree, observing that "unwritten role expectations" in scientific work environments work against women, and suggests that "a higher degree of explicitness and formality may be necessary" (p. 57) for women to succeed. Robinson and McIlwee (1991) illustrate with the differences between the aerospace industry (formal) and "high-tech firms" (informal). Engineers in the former have less power than in the latter, and the occupational status profile of women is stronger in aerospace firms than in computer software firms, for example.

Kunda's (1992) ethnographic study of a high-tech firm provides considerable evidence from the setting and "rituals" of the kind of engineering workplace that women may find difficult.

Technical exchanges take place in the men's room, and when not part of a crescendo of tech-speak, the language of presentations offers corporate products as "weapons" to "cannibalize" the competition, and solutions that are "better than sex." In open workspaces with an appearance of decentralized power, teams compete fiercely to be at the center of the organization, and, in exchanges that seem to be informal, deeply felt affirmations of individual enterprise, a conforming ideology, and sets of power relationships are codified. Management needs to keep most of their engineers "on the technical track" in a happy, productive laboratory ("an engineer's sandbox") where their creativity and technical prowess rules (or so the engineers think, even when they understand "the bullshit that comes from above"), a place described as a "boy's world" (Kunda, 1992, pp. 68, 158).

Women engineering *students*, as McIlwee and Robinson (1992) point out, succeed more in the classroom than in the laboratory, where the "vocabulary associated with tinkering" (p. 49) dominates, and where their previous socialization has not provided full access to that vocabulary. It is the ritual of the laboratory experience, though, that carries forward into the workplace and to the experience of junior engineers, not the abstract problem-solving of course examinations (this is also true of professions other than engineering). Felder *et al*'s (1995) observations of the behavior and attitudes of men and women in small cooperative learning groups in engineering are intriguing in their fit with some of this literature on gender behavior in engineering workplaces. Cooperative learning, like informal networks in the engineering workplace, is "a two-edged sword for women students, creating some problems for them [undercutting the independence necessary for career success, undervaluing of their contributions, reinforcing a sense of dependency on others for learning] while resolving others." (p. 159)

The history of engineering, in fact, bolsters McIlwee and Robinson's argument. The practice of engineering in the U.S. emerged from the shop culture of the first half of the 19th century (Calvert, 1976), and, through what were known as "mechanics institutes," was connected at an early point in its history with education (Hindle, 1976). This culture has often been accused of the sources of women's little ease with engineering. But in reading accounts of life in the shops, one is struck by how open and comparatively egalitarian they were for their time—even for ours. Information was shared freely, trade publications contained descriptions of new techniques and machinery, fierce competition was unknown, engineers and machinists worked together as apprentices, and a modicum of mobility—supported by the educational opportunities available through the mechanics institutes[23]—was available to machinists, metal-workers, and others (Calvert, 1976). Advancement of occupations, in terms of knowledge, status, and middle-class results for all was a consciousness that ran high in the shops and the organizations they engendered. The vision one reads in the accounts is typical of the 19th century American optimism of mind and will that was touted by deToqueville and mauled by Melville. This Arcadian state did not survive the rationalization of American industry, though the practical orientation of the shops pervaded engineering education through World War II (Seely, 1993) and its principal seating in the land-grant universities.

Equal proportions of men and women engineering graduates work in industry (National Research Council, 1994). It is reasonable to assume that the experience of women in industries where they are a distinct minority seeps down to undergraduates. The experience includes paternalistic attitudes (women must be protected from risk, can't work hours that putatively conflict with family roles), the perception that they had to work harder than men to prove themselves (Saigal, 1987), feedback that focuses on the person and not the task, gender segmentation in terms of task assignments (Epstein, 1991), and, no doubt, the effects of the language of the "sandbox" rituals described by Kunda (1992). If these experiences become the expectations of undergraduate students, and if the same types of experiences are reflected in the undergraduate environment, then women's migration from engineering programs is not surprising, no matter how well they perform.

What we can determine from the early career histories of the HS&B/So engineers, however, does not suggest that women were unhappy on the job, but, as Kunda (1992) points out, promotions in rank and pay are frequent at the level of junior engineer, with burnout or boredom occurring at later career stages. In 1992, the HS&B/So survey participants were asked a series of "job satisfaction" questions that are parallel to those asked in other longitudinal studies. As table 27 reveals, in only one category of job satisfaction was there a statistically significant difference between men and women:

Table 27.—Proportions of practicing engineers in the HS&B/So cohort who were satisfied with various aspects of their jobs at age 28/29 (1992)

Percentage Satisfied With:	Men	Women
Challenge of job	83.5	89.1
Pay	85.4	81.8
Opportunity for education/training	80.1	79.4
Job security	70.2*	59.5*
Opportunity for advancement	65.9	67.8

*$p \leq .05$
SOURCE: National Center for Education Statistics: High School & Beyond/ Sophomores.

This is hardly a portrait of dismay, though it is only one snapshot from one cohort at one moment in time.[24] Furthermore, the difference between men's and women's employment rates and underemployment rates in engineering were minuscule in the labor market of the mid-late 1980s and the male-female wage differential was lower for engineering graduates than any other SMET field except computer science (NSF, 1988, pp. 12-14). Certainly women were not discouraged by these surface characteristics of the labor market at the time.

In light of the messages that women have historically received from the engineering workplace, it might be worth noting that the traditional notion of an "engineering workplace" is becoming outmoded, as more and more engineering graduates are working either as sales engineers (i.e. spending most of their time in person-to-person customer support work), or in industries such as entertainment, financial consulting, and public policy development (Panitz, 1996). To accommodate the preparation of students for a labor market that builds on their technical knowledge but that requires more breadth, the undergraduate curriculum will start to resemble that of architecture in its Renaissance dimensions.

Such developments intensify problems of optimization that already bedevil the undergraduate engineering curriculum. That is, we seek a variety of outcomes that have to be compromised in order to produce ideal balance. We have time, cost, technical expertise, depth, communication skills, management tools, contexts, personal growth. The student who emerges is not a mathematical model. Lots of errors enter the flow. Both the student and the institution are always adjusting, but the student more than the institution—and along the way, the student may leave engineering.

Part 6—Leaving Engineering: Migration and Traffic

To understand how field attrition has been approached in engineering, it is profitable to examine notable studies that used institutional and national samples, and set each of them against the transcript-based history of the HS&B/So cohort. By playing off these studies against the HS&B/So, we might discover a better way to describe what happens to students. Because each of the examples cited is unique, we must construct parallel HS&B/So samples to replicate the results. The universes, then, are not exactly identical.

Institutional Study Analyses

Moller-Wong and Eide (1997) come closest to constructing an engineering path longitudinal study within an institution. Instead of a descriptive framework of stages and destinations, though, they employed a rhetoric of "risk," a statistical analysis based on the relationship between "risk" and "success," and a predictive argument. That is, they sought to predict who would be at risk of not completing a degree in engineering, at medium risk (completing a degree or still enrolled, but in another field), at mild risk (still enrolled in engineering at the end of five years), and low risk (completing the engineering program) by analyzing the interplay of variables in each successive semester of enrollment for a cohort established in the fall of 1990, and including 20 percent transfer students.

Compare Moller-Wong and Eide's distribution of 5-year fates against those of the HS&B/So students who completed the threshold courses (table 28). The comparison in table 28 marks both a matching 5-year clock and, for the HS&B/So, includes the full 11-year clock. In this replication, we are including those students who are still enrolled in engineering at age 29/30. The 5-year comparison shows a modest advantage for the HS&B/So students in degree-completion rates. The lower-rates of degree completion reported by Moller-Wong and Eide, I am sure, reflect the judgment of an individual institution that students no longer enrolled in that school are "permanent drop-outs." The comparison clearly reflects the critical factors of time and multi-institutional attendance. If, as table 8 demonstrated, the mean time to a bachelor's degree in engineering was roughly five *calendar* years, then five academic years is too soon to close the record books, and, as soon as the temporal boundary is released, there is a dramatic increase in degree completion rates. I would not be surprised if we came back to Moller-Wong and Eide's sample in the year 2000 and found that half of their "permanent drop-outs" had finished degrees in other institutions.

Moller-Wong and Eide are wise to point out that "attrition" from engineering is not necessarily an indicator of student failure. In fact, it may be part of a story of student "success." They pointed to the case of a student with the highest predicted probability of graduating in engineering, but who wound up leaving the institution and graduating from another school in geology. Such cases can be multiplied throughout cohort histories. In fact, a few institutional studies have the grace to acknowledge that overall graduation rates are more important than program graduation rates (e.g. Baker, 1988), but rarely acknowledge what our data demonstrate: that students who reach the threshold of the engineering path are more likely to complete bachelor's degrees in any field than others.

Table 28.—Five-year versus 11-year fates of engineering path students in two studies: institutional and national

	Moller-Wong/Eide (1990–1995)	HS&B/So (1982–1987)	HS&B/So (1982–1993)
Completed engineering degree	32%	39%	57%
Completed degree in other fields	17	19	26
Still enrolled in engineering	13	19	3
Still enrolled, but in other fields	8	8	2
Permanent drop-out	30	15	12*

* Includes those who earned associate's degrees, but were not enrolled at age 30.

Another institutional study that includes some of the same factors we have considered in structuring the engineering path in the HS&B/So database is Humphreys and Freeland's (1992) account of three entering freshman cohorts (1985, 1986, and 1987) in the college of engineering at the University of California at Berkeley. By one commonly-recognized measure, this is a very elite group of students (the composite mean SAT score of the *dropouts* was 1276), and one suspects that their history is not likely to be typical. But in some ways, the patterns of field attrition over four years following entrance match those of the much more diverse HS&B/So cohort, for example, women were more prone to shift to other majors than men (22 percent to 14 percent). As is the case for all institutional studies, there is no indication whether those who left the institution (21 percent of the men and 19 percent of the women) went to school anywhere else or whether they completed a degree, let alone one in engineering.

While the 4-year tracking boundary is not equivalent to degree completion, let us assume for a moment that it is, and compare Humphreys and Freeland's elite completion rates with national rates. To form a comparable HS&B cohort, we censor the "reference date" by which a student has either earned an engineering degree or not, exclude transfers from community colleges, require a minimum of 30 earned credits from 4-year institutions and do not allow attendance at any institution other than a 4-year school. More critically, to approximate the likely universe at an engineering school such as Berkeley, we also include only those students whose secondary school backgrounds (as measured by academic intensity of high school curriculum, class rank/academic grade point average, and senior test scores) put them in the top 40 percent of those who continued their education after high school, and who subsequently *crossed* the threshold of the engineering path.

Table 29.—Four-year engineering degree completion rates among students with very strong academic backgrounds

	Humphreys and Freeland (1992)	High School & Beyond/Sophomores Who Crossed the Threshold (1982–1986)
All	63.7%	63.7%
Men	64.9	64.6
Women	59.4	61.9

NOTE: The N for Humphreys and Freeland's sample was 1,232; the weighted N for the HS&B/So sample is 36,220.

What we learn from this case is that, among the most elite of an already high quality group, the degree completion gap in engineering between men and women is negligible. As soon as one lifts the restrictions imposed for this comparison on the HS&B/So sample, a 20-point hole opens (see table 4). The principal conclusions of this short report were that "students who achieve well academically both in high school and in the College of Engineering at the freshman level may choose" to switch to other majors nonetheless, that individual precursor variables are poor predictors of subsequent behavior of this elite group of students, and that even discriminant models of analysis have a pitifully low explanatory power (p. 4). As we saw in the case of the correlations between levels of high school mathematics and overall bachelor's degree completion, with every filter in the selection of the population, the explanatory power of the variable fell. When populations are selected at the right tail of achievement, very few of the input measures discussed in part 4 of this monograph can be squeezed to explain nuances in student choice and field attrition.

Alternative National Stories

The principal national stories are those of the 1989 follow-up to the 1985 CIRP survey group (Astin, 1993; Astin and Astin, 1993), a more limited 1994 follow-up to the 1985 CIRP group concerned principally with predicting institutional graduation rates (Astin, Tsui, and Avalos, 1996), the 1991 follow-up to the 1987 CIRP survey group (Seymour and Hewitt, 1997), the five-year studies of entering cohorts of 1988 and 1990 in public, research universities (Kroc, Howard, Hull and Woodard, 1997), and those derived from Graduate Record Examination test-taking groups (Grandy, 1995; Grandy, 1994).

The most intriguing of these, for purposes of understanding undergraduate field attrition, are those of Astin and Astin (1993) and Seymour and Hewitt (1997). Even though it is not based on transcript records, Astin and Astin's study for the National Science Foundation develops and demonstrates an indispensable model of the dynamics of inter-major "traffic" on which one can build. At the same time, this model illustrates the hazards of aggregation, the problems of excluding students who begin higher education in community colleges, and using a definition of "science" that is excessively broad (the Kroc, Howard, Hull and Woodard study evidences even more hazards in these matters). Drawing on the same survey series, Seymour and Hewitt (1997) try to correct the definition of "science" with some disaggregation, but then aggregate non-science fields in ways that render consistent comparison difficult. Thus, table 30 compares the recruiting/defection rates of Astin and Astin's sample for 1985–1989 against the HS&B/So rates of 1982–1993. There are three large differences in this comparison, but they are not fatal. The first is that the HS&B/So question about anticipated college major was asked in the spring of the senior year of high school, not in the fall of the first year of college. Second, the time frame for the HS&B/So cohort, 11 years, is more than sufficient to cover virtually all bachelor's degree completions, whereas that for the 1985–1989 CIRP sample is hardly sufficient. Third, Astin & Astin aggregate their major fields into six large categories, one of which covers all "non-science," and this "non-science" category includes computer science, architecture, nursing, pharmacy, allied health, and agriculture—all of which, in different ways, are applied science fields. At the same time, they include psychology and the social sciences (anthropology, economics, sociology, geography, political science) among the "sciences" (Seymour and Hewitt, on the other hand, include these under a category they call "humanities and arts").

68

Table 30.—Comparison of selected inter-major "traffic" in the CIRP (1985–1989) and HS&B/So (1982–1993) cohorts

	Initial Major Field Intention							
	Undecided		Life Sci		Phys Sci		Engin	
	CIRP (8%)	HS&B (5%)	CIRP (12%)	HS&B (3%)	CIRP (6%)	HS&B (3%)	CIRP (11%)	HS&B (14%)
Field of "Final" Major								
Business	-	27.3%*	-	6.6%	-	2.6%	-	14.1%*
Education	-	9.2	-	0.0	-	1.0	-	0.9
Humanities and Arts	-	8.6	-	4.1	-	12.5	-	5.7
Computer Sci	-	2.2	-	0.6	-	4.4	-	5.7
Communications	-	12.1	-	5.6	-	0.8	-	5.6
Health Sci/Servs	-	2.4	-	0.8	-	4.3	-	1.6
Other "Non-Sci"	-	14.5	-	12.9	-	2.5	-	6.0
SUB-TOTAL:	**68.0**	**76.3**	**42.5**	**30.6***	**46.2**	**28.1**	**39.9**	**39.6**
Social Science	15.2	14.2*	8.3	20.6	6.8	13.4	5.2	6.6
Psychology	7.2	1.0	6.5	1.6	2.6	1.6	1.3	0.4
Life Science	4.3	2.7	36.3	41.6*	4.1	18.0	1.9	0.9
Physical Sci	4.2	3.9	5.4	4.1	35.2	27.9	7.6	8.6
Engineering	1.1	1.9	1.0	1.5	5.0	11.0	43.9	43.9*

NOTES: (1) The contents of the aggregate fields in this table were determined by the definitions used in Astin & Astin (1993), e.g. mathematics is included with physical science; majors indicated by "-" are not disaggregated. (2) The HS&B/So universe was constructed to match that of Astin & Astin as closely as possible. Only those students whose true institution of first attendance was a 4-year college, who answered the question on planned major, and whose records indicated receipt of a bachelor's degree by 1993 are included. Weighted N=631,763. (3) Columns may not add to 100% due to rounding. (4) Judgments of statistical significance were computed only for the HS&B/So data and only for pairs asterisked in the columns: $*p \leq .05$.

Despite these differences, as table 30 demonstrates, if one configures the HS&B/So database to match all the filters, universes and definitions in the CIRP data base, there is a remarkable degree of agreement between the two national measures in certain key observations, particularly those involving engineering. For example, the proportion of those whose initial intention was to major in engineering, but who switched to "non-science" fields was almost the same in both studies (39+ percent) and the proportion of the same group who completed degrees in engineering was exactly the same (43.9 percent). And both studies reveal that people who abandon engineering do *not* run to the life sciences or psychology as an alternative. On the other hand, both studies show that the only significant recruitment to engineering comes from students who initially intended to major in the physical sciences (though the original physical science group is so small that there is no statistical significance in the size of the sub-group moving into engineering). The studies also differ as to the "balance of trade," so to speak, between engineering and the physical sciences. Astin and Astin (1993) tend to attribute this interchange to institutional and peer environments, whereas I hold that this particular inter-major traffic, as well as that between engineering and computer science, would be impossible without curricular momentum.

To demonstrate the case, let us release the boundaries we set on the HS&B/So data in order to compare them with Astin & Astin's calculations in table 30, go back to the empirical threshold of our sample, ask where the migrants who completed bachelor's degrees went, and give them until age 30 to get there. Table 31 presents the destinations of these migrants. Both the CIRP (Astin and Astin, 1993; Seymour and Hewitt, 1997) and HS&B/So data support the conclusion that attrition across all major categories of science and engineering is not uniform, and that the traffic flows in directions that cannot always be predicted by curricular momentum. But the engineering story is different. Among the HS&B/So migrants in table 31, 48.7 percent of men and 43.8 percent of women moved to either the physical sciences or computer science, with women weighted toward the former and men toward the latter. The only other disciplinary area to claim a measurable chunk of both male and female migrants was business. The quantitative grounds for all three of these fields argue for the momentum hypothesis to explain the 70 percent incoming traffic flow—and 60 percent of the female traffic flow—from those who left the engineering path. If one has made the curricular investments in statistics, accounting, computer programming, and physics we observed of the migrants in tables 12 and 13, it is easier to finish degrees in fields to which those subjects are constitutive.

At the same time, however, changing student perceptions of the labor market and academic fashion may alter the pattern of traffic. In light of the traffic data for the 1987–1991 period reported by Seymour and Hewitt (1997), the computer science figures are the most telling in this regard. For our sample from the mid-1980s, 27 percent of the migrants went into computer science; for the CIRP graduates of the early 1990s, the figure had dropped to 17.5 percent, reflecting an overall decline in the computer science share of graduates during that period.

As Moller-Wong and Eide point out, the migration patterns are not necessarily "losses" for *students*. They are part of the process by which learners discover who they are, where they think they will fit comfortably in the world, and acquire the knowledge and skills necessary to lead productive lives. If they happen to start in engineering, pass the threshold with courses in calculus, engineering design, mechanics, industrial engineering, and materials engineering and move into finance or geology, they will bring to their "new" disciplines both knowledge and ways of analysis that can change the boundaries of their learning and, ultimately, their jobs. Engineering, as a field, "loses" by this only when there is a projected shortage of engineers. While there is some debate about supply as a function of student choice (and not merely demographic models projected from past trends) and the extent of decline in demand, a critical shortage does not appear to be looming (Braddock, 1992).

Table 31.—Major fields of those who left engineering but completed bachelor's degrees, age 30, by gender

Percent who completed degrees in:	Men		Women		All	
Computer science	31.4	(4.27)	14.3	(7.58)	27.1	(3.67)
Business	25.4	(4.32)	14.9	(6.99)	22.7	(4.20)
Physical sciences	17.3	(3.01)	29.5	(4.59)	20.4	(4.76)
Social sciences	11.4	(3.55)	--		10.9	(2.85)
Life sciences	--		15.1	(6.93)	6.2	(1.75)
All other	11.3	(4.43)	17.0	(9.27)	12.7	(3.74)

NOTES: (1) --Too few cases to provide a reliable estimate. (2) Columns will not add to 100.0%. (3) Standard errors of the estimates are in parentheses. (4) Universe includes all students in the HS&B/So who reached at least the threshold of the engineering path and finished bachelor's degrees in fields other than engineering. Weighted N=39k.
SOURCE: National Center for Education Statistics: High School & Beyond/Sophomores.

Summary of Attrition and Migration

The statistics are a whirlwind. Every study posits a different kind of universe, a different metric, a different time period, a different methodology, and a different outcome as dependent variable. Based on the studies we have mentioned above, if we were to ask for the "retention rate" in engineering, and defined "retention" as the combination of degree completion in engineering plus students still enrolled at the end of the measurement period, the results range from 44 percent in Astin and Astin, 45 percent for Moller-Wong and Eide, 49 percent in Kroc, Howard, Hull, and Woodard; 51-62 percent for Seymour and Hewitt (depending on how one defines changing major to one "in the same group"), 64 percent for Humphreys and Freeland, and 59 percent for our own HS&B/So. That's quite a spread!

This monograph argues that a long-term, transcript-based analysis has greater validity than others, particularly in the matter of defining engineering students in terms of reaching curricular thresholds, and allowing community college transfers into the mix. What we know at the bottom line is that women constitute one out of seven students who reach the threshold of the engineering path, and one out of ten students who earn a degree in engineering by age 30. No matter how and where one draws the boundaries of analysis, there is a persistent 20 percent gap between men's and women's field completion rates in engineering (see table 4, page 16 above)[25]. The data on intent-to-major tells us that this combination of low entry level and completion gap is unhappily rare (see table 1).

Part 7—Experiencing Engineering:
Classroom Environments, Credit Loads, and Grades

It is at this point that a consideration of the effects of institutional environments and engineering program presentation, course environment, and teaching styles enters. Assume for a moment (and contrary to what is demonstrated in table 5) that everybody except community college transfer students stays in their institution of first attendance. Also assume that everybody goes through a creative freshman engineering design course and passes the threshold of the engineering curriculum. What is it that women and men experience equally and differentially?

Classroom Environments and Behaviors

Our first consideration is that of the gender composition of courses in the engineering curriculum that cut across sub-fields, comparing them to a group of courses in the physical sciences and computer science that are above entry level. This is an environmental issue. If you are a majority of students in higher education, as women are, when you look around a class, do you see a reasonable proportion of people who look like you—and how much of a difference does this make? Is the issue superficial or deep? Do men have similar experiences in elementary education or nutritional service classes? Astin (1993) would attribute the momentum generated by these experiences to "peer environment." Table 32 would suggest that for the women of the HS&B/So cohort, chemistry and (to a slightly lesser extent) computer science provided the comfort level of peer environment far better than either engineering or physics. For many women, though, it is still a shock to the system to find oneself a sudden minority (Seymour and Hewitt, 1997). But the issue turns out to be more subtle, and turns back on the differences between engineering and basic science.

Grandy (1994) is among the few to understand this distinction. She found that gender differences in self-assessment of problem-solving skills, study skills, and interpersonal skills, in student assessments of the difficulty of coursework and quality of instruction, and in self-image as a future engineer/scientist were greater among engineering students than among other SMET majors. For example, female engineering degree recipients found their courses more difficult and less enjoyable than did men. While women rated their study skills higher than did men, the opposite was true in self-assessment of problem-solving skills and in self-image as future engineers (for a partial confirmation, see Felder *et al*, 1995). In terms of ideal future jobs, women engineering students expressed stronger preferences for practical and applied work and for working with people (as opposed to things). Engineers work with both, of course: they have to know clients and culture, but the licensing criteria they face are focused far more (for example, in civil engineering) on internal forces in trusses, shear, and ground acceleration in earthquakes (Chelapati, 1990). Grandy's universe consists of students taking the GRE and continuing to graduate school, so it is a self-selected group, but ought to know better what the practice of engineering is about.

Table 32.—Gender composition of selected courses in engineering and other science curricula

Engineering Courses	Percent Female	PhysSci and Computer Sci	Percent Female
Industrial Engineering	23	Analytic Chemistry	53
Computer Applications in Engineering/Science	17	Organic Chemistry	51
		Physical Chemistry	43
Engineering Co-op	15	Computer Systems Design	41
Materials Engineering	14	Computer Organization	
Engineering Design	13	and Architecture	37
Statics/Dynamics/Mechanics	10	Thermodynamics	17
Engineering Mathematics	9	Electricity and Magnetism:	
		Intermediate Course	15

NOTE: For each course, the universe consists of all students who enrolled in the course. Students who never reached the threshold of the engineering path are included with students who reached only the threshold. The weighted universe is different for each course.
SOURCE: National Center for Education Statistics: High School & Beyond/Sophomores.

In their 1992 survey of 400 undergraduate engineering students at the University of California at Davis, Henes, Bland, Darby and McDonald (1995) asked some questions that were analogous to those used by Grandy, except of in-process students. Their answers—and the analysis—demonstrate the extent of imputation on the causes of attrition. For example, while the U/C Davis women were just as likely as men to agree strongly with the statements that "all faculty members treat me with fairness and respect" and that "I am comfortable approaching professors for help outside of class," they were less comfortable about asking questions in the classroom and had less confidence that they would finish an engineering degree. Some 30 percent of the women in this institutional survey (versus 15-18 percent of the men) expressed the most dispiriting positions on classroom participation and probable degree completion in the field. These attitudes no doubt play a significant role in attrition, but the authors here (and elsewhere in the literature) have to speculate on the connections, and some of the speculations strain credulity.

For example, Henes, Bland, Darby and McDonald (1995) claim that women in the basic science courses required of engineering majors have difficulty because the examples used "to make the material relevant are not drawn from women's prior experience." (p. 61) Their example of the examples is "the trajectory of a football." Would the problem be solved by citing the trajectory of a soccer ball, tennis ball, or frisbee instead? Yes, a football is of a different shape, but women at a selective public university who are declared engineering majors and registered for a college physics course are not the types to abandon all hope because an example is drawn from a mass culture sport. Footballs, even aggregated analogous cases, do not account for a 40 percent attrition rate in an academic field. What may be more relevant are classroom participation experiences, isolation and feelings of inadequacy in laboratories involving mechanical and electrical equipment women are far less likely than men to have seen during their high school years, and professorial ignorance of how to overcome these problems.

In recent years, study groups and teams have become more common in undergraduate engineering education, and students share and mentor each other. Take, for example, the kind of problem that is now posed in the freshman design course, e.g. analyze and redesign a system for recycling cardboard boxes in a supermarket, involve interviewing, observing the entire people-system of delivery, sorting, stocking, shelving, and disposal in a particular facility, noting paths and barriers, following recyclable material from the supermarket to a cardboard manufacturing plant, with all the machinery and chemical processes involved in recycling. You cannot do this alone. However much women participate in the culture of this process, McIlwee and Robinson report, they feel "isolated and observed" (p. 58) and often retreat from the more informal interactions of these groups, and, as Seymour and Hewitt (1997) would add, wonder whether they "belonged" (p. 242). But after the freshman design course, the engineering curriculum reverts to what Becher (1989) called a "rural" mode: students work largely in isolation, oriented toward next Monday's problem quiz with its predictable, mechanical steps and predetermined outcome.

In this respect, the recent literature on student's assessment of classroom climates shows greater variance by discipline than by gender (Constantinople, Cornelius, and Gray, 1988). In a promising study of gender-atypical fields utilizing a modified version of the Campus Environment Survey that was designed "to measure students' perception of differential treatment of men and women college students," Serex (1997) found that "there was no significant interaction between gender and academic discipline" (p. 11) in engineering, nursing, education, or accounting, but that engineering and accounting students judged the classroom climate less favorably than students in the other two fields. On a five-point Likert scale, the mean rating of classroom climate in engineering was 3.736 (S.D. = .441) for women and 3.766 (S.D. = .302) for men. These are almost identical ratings.

Credit Load

One of the common complaints of engineering students concerns what they perceive to be an excessive investment of time in classroom, laboratory and "homework" relative to that of

Table 33.—Mean calendar year credit loads for bachelor's degree completers by selected broad major fields and sex

	Mean Annual Credit Load	S.D.	s.e.	Percentage Who Stopped Out
ALL	30.3	6.5	.011	14.4
Men	29.8	6.4	.015	15.6
Women	30.7	6.5	.015	13.2
Engineering/Arch/ETech	31.3	6.1	.034	12.5
Men	31.0	6.1	.036	13.5
Women	33.8	5.3	.090	4.6
Physical Science	31.5	6.5	.069	10.5
Men	30.5	6.7	.086	13.8
Women	34.5	4.5	.103	2.7
Computer Sci/Mathematics	31.2	6.1	.042	15.6
Men	30.4	6.1	.057	21.3
Women	32.3	5.8	.061	8.2
Life Sciences	31.8	5.5	.038	8.7
Men	32.3	6.3	.064	8.9
Women	31.4	4.7	.045	8.6
Health Sciences/Services	30.4	7.0	.046	15.7
Men	28.4	7.3	.118	19.6
Women	31.0	6.9	.051	14.9
Humanities	30.7	6.5	.043	14.0
Men	29.3	6.6	.070	15.4
Women	31.5	6.3	.053	13.2
Fine/Performing Arts	30.9	6.7	.052	9.3
Men	29.7	6.6	.081	15.9
Women	31.6	6.6	.066	4.9
Social Sciences	30.2	6.5	.027	16.4
Men	30.0	6.1	.038	16.2
Women	30.3	6.7	.039	16.6
Business	29.5	6.2	.021	13.2
Men	29.0	5.7	.026	12.2
Women	30.2	6.7	.032	14.3
Education	29.7	6.8	.044	19.2
Men	25.8	5.9	.090	36.9
Women	30.5	6.7	.049	15.1

NOTES: (1) Standard errors (s.e.) have been adjusted for design effects. (2) Universe includes all HS&B/So students who completed bachelor's degrees by 1993, whose undergraduate records were complete and for whom true time to bachelor's degree could be computed. Weighted N=824k.
SOURCE: National Center for Education Statistics: High School & Beyond/Sophomores.

their peers majoring in other fields, or, as Seymour and Hewitt (1997) refer to it, "curriculum overload." If credits are an accurate proxy measure for time, and if the credit-values of engineering courses, in particular, match the investment of student time, then we should see these student complaints manifest in average annual credit loads. Table 33 presents the average annual credit loads, by major, for those who completed bachelor's degrees. The formula aggregates all credits earned between the true date of first attendance (see p. 22 above) and the date of the bachelor's degree, and divides by the *calendar time* from the true date of first attendance to that of the degree (this is elapsed time, not enrolled time). On an *academic year* basis, then, these figures are slightly deflated because the calendar year includes summer terms and, where applicable, special January terms, and also by the proportion of students who stopped out of college at some time (the "non-continuous enrollment" percentage)[26].

What do we see in table 33? First, in every major academic field except the Life Sciences and the Social Sciences (where the differences are statistically insignificant) women carry a higher annual credit load than do men, partly because women were far less likely to stop-out of college, and that the variances (standard deviations) in core SMET fields are consistently lower for women than men. That means that women in SMET majors are behaving more similarly than are men. Second, students in all core SMET fields carry an average of about one credit more per year than students in non-SMET fields. Again, this comparison is influenced, in part, by stop-out behavior, as only one non-SMET field (Fine and Performing Arts) evidences low stop-out rates, whereas three SMET fields—engineering, physical sciences and life sciences—exhibit stop-out rates significantly below the mean. Third, there are no differences in mean annual credit loads among all the core SMET fields, i.e. engineering students are no more burdened by credit load than, for example, biology majors or computer science majors. Of the engineering path groups, as table 34 demonstrates, the migrants "succeeded" in lowering their credit loads only if they changed majors to business:

Table 34.—Mean calendar year credit loads for students on the engineering path

	Mean Annual Credit Load	S.D.	s.e.		Percent Who Stopped Out	
Threshold only	30.5	7.5	.093	‖	16.7	(9.08)
Migrants	29.7	6.7	.078	‖	12.1	(4.73)
to business	26.6	5.7	.125	‖	--	
to physical sci	31.7	7.5	.246	‖	--	
to computer sci	31.3	6.6	.132	‖	--	
Completers: terminal	31.0	6.2	.041	‖	12.7	(2.81)
Completers: continued	32.1	5.7	.059	‖	11.0	(4.52)

NOTES: (1) Universe includes all students who reached at least the threshold of the engineering path, excluding those in 2-year only engineering tech programs. Weighted N=139.5k; (2)--N insufficient to yield an estimate.
SOURCE: National Center for Education Statistics: High School & Beyond/Sophomores.

It could be, of course, that engineering students feel that the amount of work they do is not adequately reflected in the credit-accounts. This perception may be part of what Seymour and Hewitt (1997) describe as the "weed out" system, an environment characterized by feelings of competition among students, endless laboratory hours with few credits attached, and lower-than-expected grades[27]. The "weed out" system is a phenomenological construction of students, and for the men who leave engineering, as Seymour and Hewitt point out, interpretations become justifications.

Academic Performance and the Grades of Men and Women

Paradoxically, one of the reasons men appear to be carrying lighter credit loads than women is that they do not perform as well academically, hence are often repeating courses (Felder *et al*, 1995 made a similar observation of men in their longitudinal study of chemical engineering students), and spending more calendar time in undergraduate education. A key guide to this phenomenon is the ratio of credits earned to credits attempted. The difference between the two is due not merely to failures, but also to withdrawals, incompletes, and "no credit repeats" in colleges that allow students to repeat a course for a better grade.

For all students who earned more than 10 credits, whether or not they eventually earned degrees, 25 percent of the men versus 19 percent of the women earned less than 90 percent of the credits they attempted (t=3.12). Along the engineering path, the greatest weaknesses in this respect were, not surprisingly, among the migrants:

Table 35.—Proportions of HS&B/So students earning less than 90 percent of credits attempted, by engineering path status

Earning Less Than 90% of Credits Attempted

	Men	Women	All
No engineering path	26.8*	19.5*	22.6
Threshold only	24.4	22.2	24.1
Migrants	39.7*	27.2*	36.8
Completers	5.2	5.6	5.3

NOTES: (1) The universe consists of all students who earned more than 10 credits and for whom a credit ratio could be calculated. Weighted N=1.96M. (2) *p≤.05.

SOURCE: National Center for Education Statistics: High School & Beyond/ Sophomores.

Of course completers will be more academically successful than groups including non-completers, but the migrant group had a higher bachelor's degree completion rate (65 percent) than the threshold only group (61 percent), let alone the group that did not engage the engineering path (43 percent). The migrants obviously encountered academic problems along the path and even among those who completed degrees, GPAs were lower. Seymour and Hewitt's interview subjects back up this observation, though there is an obvious difference in the range of grades:

	Seymour & Hewitt (self-reported GPA)	HS&B (transcript GPA)
Completers	3.5	2.91
Migrants	2.85	2.53

We know that students' estimates of GPA are always inflated, and these differences (self-reported v. transcript-based) are considerable. Interestingly enough, Seymour and Hewitt also report that 34 percent of their overall SMET sample left for other majors due, in part, to "low grades in early years," but self-reported grades for those students are not disclosed, and it would be revealing to discover what students think "low grades" mean.

Based on the CIRP surveys, Astin (1993) tells us that engineering students have lower grade point averages than others. Again, we can test this contention with the transcript data of the HS&B/So, match our universe to Astin's (that is, only students who entered 4-year colleges directly from high school, and with records censored as of the end of 1986). In table 36, we find a more complex story:

Table 36.—Mean grade point averages of engineering students who entered 4-year colleges directly from high school, and after 4.5 years of college study

	Completed Bachelor's	Did Not Complete
No Engineering Path	2.88 (SD=.51)	2.36 (SD=.60)
Threshold Only	2.99 (SD=.45)	2.04 (SD=.32)
Migrants	2.75 (SD=.49)	2.03 (SD=.44)
Completers	2.89 (SD=.55)	----

SOURCE: National Center for Education Statistics: High School & Beyond/ Sophomores.

Yes, people who initially attempt to major in engineering, but do not perform well, have lower GPAs than people who attempted to major in other fields but who did not perform well. But among completers in engineering, there is virtually no difference in GPA when compared with completers who never attempted to major in engineering.

Grades and GPA are common tools in the analysis of college student careers, and not the least because students talk about them. In local studies, grades are reported by participating faculty or obtained from registrars, hence are very reliable. On CIRP longitudinal studies or those based on the student information background questionnaires of de facto national examinations such as the SAT or GRE, they are reported by students, and never at the course level. The HS&B/So grades come from transcripts, and were standardized across 2500 institutions[28]. If we accept Warren's guidance that grades "are sufficient for the limited purposes for which they are intended, which [include] . . . . acceptable completion of a segment of an educational program," and are fairly reliable (Warren, 1989, p. 68), we will not place undue emphasis on them. But they are prominent in student consciousness, particularly when the issue is what I would call "the cusp of an A."

Table 37 displays a standardized distribution of grades for men and women in nine (9) engineering course categories. Only categories with weighted Ns of 10,000 or more for both men and women were included. A few words about the technical aspects of this table are necessary. First, the particular file one must use to obtain letter grade distributions, while weighted only for students with complete undergraduate transcript records, is such that I cannot claim statistical significance for a specific match of letter grade by course for men and women. To ascribe significance, for example, to the fact that 7.1 percent of men but 13.1 percent of women received "Ds" in chemical engineering courses is not warranted. Second, three of the grade categories are umbrellas:

- "F" includes all penalty grades;

- "W" includes withdrawals (but not drops), no-credit repeats, and unresolved incompletes; and

- "Pass" includes all cases in which a student received credit without a standard letter grade.

Third, the percentages on the rows may not add to 100 percent because there is an additional category of letter grade on the file used for this analysis, a category that includes courses that carried neither credits nor grades, drops, examination entries that did not carry credits, and other institutionally idiosyncratic notations.

With this background in mind, we can examine the general distribution of grades by gender. Women appear to perform better than men in mechanical engineering, industrial engineering and engineering mathematics. Men appear to perform better than women in computer engineering. Otherwise, the differences are comparatively minor. The distribution in the key category of Statics & Dynamics, a category crossed by students in nearly all engineering

Table 37.—Grades of men and women in selected engineering course categories

	As	Bs	Cs	Ds	Fs	Ws	PASS
Intro Engineering							
Men	32.6%	29.0%	15.4%	3.7%	3.9%	1.9%	11.6%
Women	12.6	46.7	21.0	9.5	Low N	Low N	Low N
Statics,Dynamics							
Men	21.0	27.5	29.5	9.2	7.1	4.7	0.9
Women	18.6	34.2	24.5	5.7	7.6	7.6	Low N
Chemical Engin							
Men	20.7	28.5	25.1	7.1	Low N	3.6	10.2
Women	21.6	24.5	26.3	13.1	11.8	Low N	Low N
Electrical Engin							
Men	29.5	29.1	22.0	8.0	5.3	4.5	1.0
Women	21.6	40.4	22.8	8.4	Low N	Low N	Low N
Computer Engin							
Men	33.6	32.7	18.2	7.6	3.4	3.2	Low N
Women	17.4	31.6	31.1	Low N	5.9	8.2	Low N
Mechanical Engin							
Men	23.5	35.3	26.8	6.5	2.4	3.7	1.2
Women	32.6	39.3	17.3	5.9	Low N	Low N	Low N
Industrial Engin							
Men	26.1	37.4	27.8	Low N	Low N	Low N	3.0
Women	43.2	37.1	15.5	Low N	Low N	Low N	Low N
Materials Engin							
Men	23.9	35.0	24.8	7.4	5.6	Low N	Low N
Women	30.8	23.5	25.2	11.9	Low N	Low N	Low N
Engineering Math							
Men	16.6	32.1	30.5	8.2	6.0	6.0	Low N
Women	36.2	34.5	20.6	Low N	Low N	Low N	Low N

NOTE: Rows will not add to 100.0 percent due to low-N cells.
SOURCE: National Center for Education Statistics: High School & Beyond/Sophomores.

specialties, is a model indicator: 57-59 percent of the grades for both men and women were Bs and Cs, 21 percent were Ds, Fs, and Ws for both. For those concerned about "the cusp of an A," the score for the nine courses is a draw: a higher percentage of women received As in four; men "won" four; and one (chemical engineering) was a tie.

Other similar observations can be found in the literature, e.g. Felder, *et al* (1993) found no difference between women's and men's passing rates (with C or better) in the sophomore introduction to chemical engineering. Grades, Felder *et al* found, were far more influenced by expectations, perceptions of obstacles to academic success, amount of time spent working, participating in extracurricular activities, and socializing. These are far more common-sense independent variables than gender.

The bulk of women who leave engineering—and men as well—are not going to leave because of grade distributions such as these. Yes, we do have a group that crosses the threshold, does not perform well, and leaves the engineering path. But this group accounts for only 8.5 percent of all students in 4-year college attendance patterns who reached the threshold (see table 2). Seymour and Hewitt's interviews help us account for the others who leave. They teach us that among those who exit SMET fields for reasons other than poor academic performance, the most compelling factors in their decisions are:

Table 38.—Factors in the decisions of students who switch from science or engineering majors to other fields, according to Seymour and Hewitt (1997)

	All SMET Students	Engineering Students
Percent of "switchers" citing:		
Poor teaching by SMET faculty	90	98
Reasons for choice of SMET major prove inappropriate	82	94
Inadequate advising	75	81
Loss of interest in SMET	60	66
Non-SMET major offers better education/more interest	58	57
Curriculum overload	45	55

SOURCE: Seymour & Hewitt, 1997, pp. 33, 46.

This configuration conveys two messages: (1) teaching and advisement in SMET fields are far below par, and, in engineering programs, worse; and (2) entering students do not know what they don't know: they have images of a future in which they discover dissonance, and seek to resolve that dissonance into harmony. The former message is telling; the latter is not—it happens in every field. In Seymour and Hewitt's interviews, both men and women who leave SMET majors equally regard loss of interest ("turned off on science") and poor teaching to be major factors in their decisions. The teaching issue is a serious one and is within our power to change. Astin (1977, 1993) has often demonstrated that frequent student interaction with faculty has positive effects on student development, involvement, and retention. On closer examination, Astin and Astin (1993) found that this conventional wisdom did not hold for engineering students. In a devastatingly reserved speculation, they note that "greater interaction with faculty may not have the same positive effect on engineering students simply because these interactions are less likely to be perceived as favorable" (p. 4-28). The students who abandon the field with the bad taste of poor instruction may not be very satisfied with their higher education, even if they finish degrees.

The Bottom Line of Satisfaction

At the end of it all, were the HS&B/So engineering path students satisfied with their undergraduate experience? Astin (1993) claims that, among other negative consequences of majoring in engineering is a greater degree of dissatisfaction with higher education. Again, it is fortunate that we can test this hypothesis against a different portrait of "engineering students." In 1986, the members of the HS&B/So cohort were asked about their degree of satisfaction with various aspects of their higher education experience. The topics can be aggregated into four categories (academic aspects of college, non-academic/institutional environment, preparation for work/careers, and costs). In turn, one can build a composite index of satisfaction from the four categories of responses. Including only those students who earned more than 30 credits from 4-year colleges as of 1986, the composite index shows no differences in the proportions of students indicating moderate to severe dissatisfaction with their college experience; but when the neutral value on a 5-point scale is included, a bi-modal pattern involving threshold and migrant students on the engineering path emerges:

Table 39.—Composite index of dissatisfaction with higher education of students on the engineering path and others

	Proportion Dissatisfied	Proportion Neutral to Dissatisfied
No Engineering Path	11.8%	29.7%
Threshold Only	6.4	17.2
Migrants	11.4	39.2
Completers	10.9	31.4

NOTE: The universe consists of all students who earned more than 30 credits from 4-year colleges and answered all questions about the degree of their satisfaction with higher education in 1986. Weighted N=970k.

SOURCE: National Center for Education Statistics: High School & Beyond/ Sophomores.

The migrants displayed the highest degree of dissatisfaction with academic (38 percent) and work preparation (31 percent) aspects of their experience and the lowest degree of dissatisfaction (21 percent) with the cost of their education. The "threshold only" group had the lowest degree of dissatisfaction with work preparation (15 percent) and non-academic/institutional environment factors (16 percent).

Conclusion: the analysis of dissatisfaction is rather nuanced, and differences between any category of engineering path student and non-engineering students emerge only when, paradoxically, "dissatisfaction" is defined more generously. Even then, students who complete engineering degrees are no more dissatisfied with their college experience than non-engineering students. It may very well be, after all, that people who persist along an academic path genuinely enjoy what they are studying, and that enjoyment may override any modest dissatisfactions with costs or institutional environment.

Conclusion: What Did We Learn?
Where Do We Take the Information?

The problems that stimulated this inquiry might be stated as follows: we have not been tracking students in higher education very well in terms of initial field choice and change of major as they search for academic identity. The "traffic" among the disciplines moves at a high rate; provosts and deans worry about this because it affects their ability to plan; and some fields worry about attrition and migration because they have historically exhibited equity problems. Our "radar screen" on these problems has not been functioning well, and academic advisers do not possess enough three-dimensional information on student careers to help. I chose engineering as a case, in part because the literature has lead us to believe that its enrollments were volatile and its attrition rate very high, but also because it attracts all the variables affecting field choice, persistence and migration, and because it is offered in a small enough set of institutions so that the story line is not distorted by wild swings in institutional environments.

In engineering, too, we have a highly gender-segmented field; the movement toward selecting engineering as a career and a college major begins in secondary school, where students acquire curricular momentum; a higher proportion of women than men have that curricular momentum but do not choose to explore even the threshold of the engineering path in college; those women who begin the study of engineering in college are less likely to complete a degree in engineering than men, thus exacerbating the segmentation. This story is that of the cat chasing its own tail. Where and how does one break the cycle?

What we learned by using the high school and college transcripts of a national longitudinal study that followed a cohort from age 15 to age 30, picking up some labor market experience variables and changes in values, attitudes, and aspirations along the way, reinforces the dual-nature of this story line. By keeping our focus on the student as story-maker, we found that there are a number of paths, and they evidence different textures, contexts, and dynamics.

Student choice is an evolving phenomenon, and it is not very accurate to talk about "attrition" in any field until a student actually starts to major in it, then leaves for something else. We found that both women and men who leave the engineering path are more likely to take their curricular momentum into computer science and the physical sciences than other majors; and women who leave the engineering path are more likely to complete bachelor's degrees than are men. So how different are women's and men's behaviors along this path? And how much of a "loss"—and to whom—are we talking about?

Changing the "Bad Press"

Paradoxically, when equity questions are foremost on the table of higher education, we like to tell bad stories. When we note low participation rates of women in engineering, starting at the moments of career awareness in secondary school, we look for explanations in a variety of environmental forces. With the best of intentions, we spin out these stories in order to suggest changes in the environment. But the effects of these stories are counterproductive: they give engineering a bad rap. If only 5-6 percent of female high school seniors indicate a preference for engineering as a college major, and the proportion of female 4-year college students who test the threshold of the engineering path drops to the 3 percent range, then we have an initial problem of recruitment, and ought to look more carefully at the stories we are telling about engineering and engineering students.

Astin's (1993) analysis paints engineering as a "bad" field because even those aspects of student growth, values development, etc. that are positively affected by majoring in engineering are not wholly honorific ones, e.g. the belief that the primary value of college is to increase earnings—except that is the official propaganda line that we feed to *everybody* as the reason for going to college. As Becher (1989) noted, engineers "come across to their more hostile observers as dull, conservative, conformist, and mercenary" (p. 28). This does not sound like a very encouraging environment for women, and not a very good way to recruit them.

In light of what we know about actual practices in different kinds of engineering workplaces (McIlwee & Robinson, 1992; Whalley, 1986; Kunda, 1992; Bucciarelli and Kuhn, 1997), this negative view of engineering ought to be reexamined. There are aspects of engineering culture in some industries and firms that are characterized by creativity, light-heartedness, cooperation, and resistance to the authority of management, and, as Becher (1989) remarked of those who take a more sanguine attitude, engineers are seen as "likeable and enthusiastic; as creative, lateral thinkers; and as having a broad outlook." (p. 29) Ironically, as we have noted, this culture turns out to be a difficult one for women (McIlwee and Robinson, 1992).

The point is that one cannot draw a cardboard portrait of engineering students, nor can one paint their experience as bleak and uninviting. Yes, in some respects, those who cross the threshold of the engineering path share some academic characteristics and some non-academic values. But they are a diverse enough group to share some of the same characteristics and values with their non-engineering path peers, for example, the proportion who considered earning an MBA or the proportion evidencing general dissatisfaction with the higher education experience.

More recent research, and of a type that gets beneath the surface of survey responses, has demonstrated that on the conceptual front of broad scientifically-based problems such as global climate change, at least, there is very little difference between freshmen entering in engineering and those entering in other disciplines (Atman and Nair, 1996). They possessed a similar range of concepts and were equally as likely to attribute the problems to both social behavior and technological developments. Engineering students, however, were more confident in technology as a solution. That makes sense. That confidence is one of the reasons they are majoring in engineering!

Changing Recruitment—and More

This theme arose in the consideration of the differences between engineering and science. Engineering is not an easy field to understand, particularly by adolescents. While it has a high degree of imagibility, the initial image is not accurate. To young women, engineering appears as a good old boys club where guys tinker with machines and crack jokes about technical incompetence (Hacker, 1981), part of the "manual legacy" of engineering (Whalley and Barley, 1997). To both young women and men, engineering appears as experimental science. Neither women nor men will choose engineering for the right reasons unless both the profession and engineering educators can reach out to a broad population with a full portrait of the richness of culture and practice, with a well-defined map of its intersections with and divergences from bench science. It may be expensive, but a traveling demonstration that put clients and engineers together on high school stages to play out a project design that has cultural, economic, and political dimensions in addition to engineering tasks and calculations and scientific knowledge may do more to teach large populations of adolescents what engineering practice is about than summer camps for small populations. If presented creatively, these demonstrations of the "object world" and "social world" of engineering (Bucciarelli and Kuhn, 1997), will show that women naturally belong in engineering practice.

But would the undergraduate engineering curriculum follow through on such a vision? Would it begin with a freshman design course that included more than casual instruction in the sociology and culture of engineering practice and careers? If students who reached the "threshold" of the engineering paths *but did not cross* made their decision in full knowledge of what engineering was about, we would witness less "wastage" on the other side of the threshold.

There are further implications of a different vision of recruitment for the undergraduate curriculum. When the Board of Engineering Education of the National Research Council asks whether the current curriculum "instill[s] a sense of the social and business context and the rapidly changing, globally competitive nature of today's engineering" (Board of Engineering Education, 1994, p. 21), it asks a question with multiple dimensions. First, is it sufficient to imply that a curriculum do no more than "instill a sense"? If employers are unhappy with engineering graduates, their attitude is not likely to improve if all we do is to "instill a sense." Real stuff is called for, i.e. a bill of specifics similar to what we would ask for in dynamics of all engineering majors, regardless of sub-field. Second, if we ask for a bill of particulars, how do we deliver it? The challenge is similar to other objectives of undergraduate education that cut across majors, e.g. demonstrable knowledge of the multiplicity of cultures that comprise American society. Do we deliver "social . . . context[s]" of engineering with a separate course, or do we integrate these materials into existing courses? Do we require a course or experience in international studies, gin up a course on the worldwide status and practice of engineering, or integrate case study readings in senior seminars? These are serious challenges, and more serious, still, because engineering students face a daunting set of requirements in terms of credit-load. Engineering education faces a tension between superficial coverage ("a sense") and the additional credits and time that come with depth.

If we want to add all we want to add to the engineering program to reflect the richness of engineering practice, and include design projects and co-op, then perhaps the first engineering degree should also add a capstone design project and be a 5-year Master's degree instead of a Bachelor's. Let us also say that we will charge only 4 years' worth of tuition for it. Students will not analyze opportunity costs of the fifth year in the same way if we take this tack. Students of limited means, among whom minorities are overrepresented, will not be as daunted by undergraduate finance. Slow it down! For everybody! The workplace will teach speed, because clients will demand just-in-time. But I would rather come into the workplace with something learned deeply at the pace of care, and then speed it up, than with something learned like a skipping stone. If you blast entering might-be engineering students with loads they perceive to be 36-40 credits in their first year, you will not have many real engineering students by the second year.

Not About Engineering Only; Not About Women Only

Much of what we have learned from the engineering path students in the HS&B/So cohort applies to fields other than engineering, and to broader disciplinary aggregates than SMET. For example, Tobias (1990) would focus attention on the people who reach our threshold levels—whether in engineering or in other basic and applied sciences—and seek ways to recruit them into scientific majors and careers under the notion that scientists are made, not born. And Tobias, Chubin and Aylesworth (1995) cite, with approval, the growth of "strong minors," dual majors, and applied science programs that both "diversify and integrate" fields as promising approaches to recruitment (pp. 104, 107).

But the appeal of a broader curriculum implied in such combinations as chemistry and business or materials science and environmental studies, will not, in itself, evidence staying power without classroom-level changes. The strategy of lower-division science courses, Tobias (1990) argues, should be to cultivate, not weed out (though "weed out" may not be the most felicitous of terms to describe what happens to whom), and these courses are still part and parcel of new program combinations. No discipline can maintain enrollment shares with a weed-out system, and yet each discipline has a culture that naturally diverts some students onto other paths. To believe that fields other than science or engineering have an easier time recruiting because they offer a community of intense interest and involvement, that students are not subject to what Tobias calls the "tyranny of technique" in accounting or psychology or journalism, may be somewhat naive. Yes, we can improve the way science and engineering are taught, particularly in large institutions, but let us not pretend that these are the only domains in which such an effort is necessary.

The analytic rubrics we have applied to the particular problem of women in engineering, too, can be turned on their heads in a professional field such as nursing or an academic discipline such as psychology, both of which exhibit gender segregation on the other side. If we are at all honest about concern with structural imbalances in participation by members of any demographic group in a profession or academic field, then we ought to be asking the same questions about imagibility, choice, curricular momentum, classroom experiences, and migration in these fields that we asked of engineering. The model presented in this monograph can be used in other fields subject to the notion of a "threshold," and, provided one has sufficient numbers for analysis, for racial/ethnic minority populations, too. What are these thresholds? who explores them? who crosses them? Men have been a distinct and declining minority in undergraduate education (enrollees and degree recipients) for more than a decade. Does this decline result in more field segregation? If so, why? If we are committed to nothing more than the economic utility of more demographically balanced work forces, we owe ourselves answers to these questions.

Notes

1. The annual surveys of entering college freshmen of the Cooperative Institutional Research Program consistently show a higher percentage of "undecideds" at highly selective universities and four-year colleges. See, for example, Astin, Korn and Berz, 1990. Kroc, Howard, Hull and Woodard (1997) showed a 31.4 percent undecided rate for entering 1988 and 1990 freshmen in public, land grant, research universities, all of which are at least moderately selective.

2. In their survey of engineering institutes in 14 countries, Dorato and Abdallah found that the average proportion of women in the programs was 10 percent, with highs in the 17–20 percent range (China, South Africa, and Sweden). In the 1995 OECD survey, the mean for 21 reporting countries was 18 percent, with a high of 28 percent in Portugal.

3. The National Center for Education Statistics does not accept response rates less than 85 percent. This response rate criterion is part of all contracts for longitudinal studies surveys.

4. The primary driver of this strategy is the stratified sampling design of NCES longitudinal studies that require Taylor-series standard errors, and the software programs that utilize Taylor-series methodology will not produce either an estimate or a standard error if the unweighted N in a cell is less than 30 (see Technical Appendix). If we have five aggregate categories of academic career histories divided by five categories of race/ethnicity, we wind up with too many cells where $N < 30$, and no reliable estimates are possible.

5. The proportions of 4-year college students in the High School & Beyond/Sophomore cohort who reached at least the threshold of the engineering path and thus, as a universe, are comparable to the subjects of the Felder, Mohr, Dietz, and Baker-Ward (1994) study, by urbanicity of high school are (standard errors are in parentheses):

	Urban	Suburban	Rural
Threshold Only	17.3% (4.89)	60.1% (7.33)	22.7% (6.68)
Migrants	17.8 (4.50)	57.5 (5.98)	24.8 (5.26)
Completers	11.3 (2.31)	63.0 (4.00)	25.6 (3.65)
All Students Who Attended Any Kind of Institution	**19.4% (1.31)**	**51.7% (1.65)**	**28.9% (1.36)**

6. For example in both its 1987 and 1991 surveys, EMC data count about one-third of first year enrollments as "pre-engineering/undeclared majors," and note that some responding institutions neither estimate nor report first-year enrollments at all because, in our terms, it is not clear who will cross the threshold. Nor do the EMC data track pre-engineering transfer programs "at two- or four-year schools with no engineering programs of their own"

(Engineering Manpower Commission, 1991, p. xii), thus cannot be used to estimate national attrition.

7. We assume that high school transcripts begin in grade 9, the model year of which was 1978 for the High School & Beyond/Sophomore cohort, and we know that the college transcripts were collected in 1993, even though some students had completed or left college many years earlier.

8. After excluding 180 students whose records consisted of nothing but GED/secondary school-level and developmental courses, there are 8,215 students in the HS&B/So transcript file, of whom 91 percent had complete or probably complete records (i.e. what was missing was incidental, e.g. a summer school transcript), and another 3 percent had complete records from 1982-1987 only.

9. The "threshold" criteria consisted of dominant clusters of sources as observed on transcripts. As engineering students proceed beyond the threshold, other clusters may appear, but our coding system cannot identify them. For example, Aerospace Engineering, Theoretical Aerodynamics, Aeroelasticity, and Aerospace System Design would all be aggregated under the same code, no matter where they were offered at a particular institution. There are over 1,000 course codes in the "College Course Map" (CCM) system, of which engineering claims 30, engineering technology another 31, and architecture 8. The classification system in these fields was reviewed by different faculty panels in both 1990 and 1995. In the opinion of these panels, the aggregations were deemed to be adequate for statistical analysis in national samples of postsecondary course work by cohorts of students majoring in many fields.

10. The CIRP longitudinal studies face a severe problem of time-censoring in analyses of actual graduate school enrollment and degree attainment. Even if they covered time spans similar to those of the NCES longitudinal studies, they don't have transcripts, and the transcripts enable us to distinguish between mere post-baccalaureate course work and incomplete graduate degrees (let alone completed graduate degrees). The distinction between post-baccalaureate course taking and incomplete graduate degrees arose in the process of examining the standardized records of 8,395 students in the High School & Beyond/So. The questions addressed by the reviewers were: do the courses taken after the bachelor's degree form a coherent pattern and, with reference to standard degree requirements, can one say that the student took these courses in pursuit of a particular degree? Some cases are obvious, for example, the transcript from a law school with courses in contracts, torts, municipal corporations, wills and estates—but nothing more—taken over a period of a year. The record then goes blank. It's a case of an incomplete degree. But if we found a student who, over a period of three or four years after the bachelor's degree, took a half dozen courses in three different subjects in three different kinds of institutions, then we are looking at course taking, not a degree program.

11. For this comparison, the HS&B/So taxonomy was configured to match the CPEC categories, with some leaps of faith as to how courses were classified by the CPEC contractor nearly 20 years ago. For example, we included business economics under "economics" and corporate finance under "management."

12. Among those HS&B/So students who entered postsecondary education, the proportion completing bachelor's degrees by age 29/30, by highest level of mathematics studied in high school was:

	ALL	MEN	WOMEN	Percent of All
Calculus	82.6% (2.28)	79.7% (3.19)	86.0% (3.13)	6.9%
Pre-Calculus	76.0 (2.35)	74.7 (3.34)	77.6 (3.59)	6.3%
Trigonometry	65.5 (2.15)	65.4 (2.67)	65.6 (3.27)	12.0%
Algebra 2	46.0 (1.46)	46.7 (2.11)	45.3 (1.92)	28.3%
<Algebra 2	17.7 (.909)	17.1 (1.26)	18.2 (1.18)	46.5%

13. The odds ratio applies to all high school graduates in the HS&B/So data files for whom high school transcripts were available and who earned more than 10 postsecondary credits, and was higher for women (2.33 to 1) than for men (2.21 to 1). Paradoxically (though it is a matter of common sense), taking calculus in high school and performing well on either of the Advanced Placement calculus examinations would not result in as strong odds of earning more than 4 credits of calculus in college. It is important to note than an odds ratio describes a probability, not an empirical event.

14. Some 89 percent of U.S. secondary schools, affecting 96 percent of high school enrollments, offered chemistry in 1982. Only 36 percent of secondary schools, enrolling 44 percent of students, offered physics (West, Diodato, and Sandberg, 1984, p. 56).

15. While the high school transcript file for the High School & Beyond/Sophomore Cohort contains advanced placement test scores, they are very inconsistently recorded. So, too, is the labeling of "honors" and "AP" courses in the over 1,000 high schools from which the sample was drawn. Analysts thus often search for other proxies.

16. This is a weighted percentage for the universe of those who continued their education in any way after high school. If we used the weight for all students in grade 12, whether or not they ever graduated from high school, only 37 percent took the SAT or ACT.

17. ACT scores were converted to the SAT scale using the method developed by Astin, and following a "lumpy score" distribution such that only composite scores can be used with confidence. PSAT scores for students whose records were missing SAT or ACT scores were converted to the same lumpy scale using an internal analysis prepared in 1982 by the Educational Testing Service as part of its ongoing reviews to update tables in the "PSAT/NMSQT Interpretive Manual for Counselors & Administrators."

18. The "mini-SAT" results are reported in percentiles, not scales. The composite mean percentiles for men and women at the principal destinations of the engineering path are:

	Mean	S.D.	s.e.
Threshold			
Men	79.2	17.3	.202
Women	75.7	26.8	.763
Migrants			
Men	84.7	18.8	.202
Women	76.1	20.6	.406
Completers			
Men	82.0	17.5	.106
Women	83.0	20.5	.364

19. The income floor is a minimal filter to ensure that when a respondent tells an interviewer what he/she does for a living that there is actually a "living" to back up the claim to an occupation.

20. Wagenaar used a single year (the base year) response only for the "life values" question concerning the importance of money, and turned it into a dummy variable for use in multivariate regression analyses. While helpful in an analysis of initial occupational aspirations and planned major field of study in college, this strategy does not discriminate the contexts of pecuniary values that the HS&B/So data allow, nor does it take account of changes over time in relation to the kind of broad curricular experiences reflected in the destinations of the engineering path.

21. For example, in the case of the change in the ratio set against life values for male "migrants" between 1982 and 1986, the $t = 13.1$ where the Bonferroni adjustment requirement for a significant t at $p \leq .05$ is 2.865.

22. Women moved into the majority of degrees in pharmacy in the academic year, 1986–87 (*Digest of Education Statistics, 1990*, table 240, p. 281), and in veterinary medicine, during the academic year, 1988–89 (*Digest of Education Statistics, 1992*, table 260, p. 283).

23. The New York City Mechanics Institute, for example, was founded in 1831, inaugurated with "lectures on natural philosophy and chemistry" and "rapidly built a library, a museum of models and machines, and a membership that by 1836 reached three thousand." (Hindle, p. 103).

24. Using a more sophisticated statistical analysis and applying it to a sample of 9,350 GRE test takers in 1990–91 who had earned bachelor's degrees in SMET fields at least 5 years before taking the GRE, Grandy (1996) found that men were more likely to respond favorably about the challenge of their jobs, and women more likely to judge their job security as adequate.

25. Both Felder *et al* (1995) and Seymour and Hewitt (1997) disagree with this differential degree completion rate. They found equal proportions of men and women engineering students either completing degrees in engineering (Felder *et al*) or staying in the same engineering major (Seymour and Hewitt).

26. Astin and Astin (1993) asked students whether they had stopped out or withdrawn from college at any time, and came up with much lower figures than the transcript-based analysis shows here. Our criteria for stop-out (non-continuous enrollment) was either a single break of two or more semesters or 4 or more quarters or two or more breaks of one or more semesters or 2 or more quarters. These are very rigid criteria designed to apply to long-term records. The students in the 1989 follow-up to the 1985 CIRP freshman survey exhibit lower rates of stop-out because (a) the time period is shorter and (b) there are no community college transfer students in the CIRP universe.

27. Seymour and Hewitt ascribe this environment to all SMET fields, thus overlooking a number of confounding cross-currents. In the face of the "urban" organization of academic work in some branches of basic science (Becher, 1989), with large research teams and articles published with dozens of co-authors, for example, it is misleading to label a huge range of fields as destructively competitive.

28. In the course of editing the HS&B/So transcripts, over 700 institutions were telephoned to translate grade, credit, and course information (see Adelman, 1996).

References

Accreditation Board for Engineering and Technology (ABET). 1991. *Criteria for Accrediting Programs in Engineering in the United States.* New York: Author.

Adelman, C. 1994. *Lessons of a Generation: Education and Work in the Lives of the High School Class of 1972.* San Francisco: Jossey-Bass.

Adelman, C. 1995. *The New College Course Map and Transcript Files: Changes in Course-Taking and Achievement, 1972-1993.* Washington, DC: U.S. Department of Education.

Adelman, C. 1997. *Leading, Concurrent or Lagging: the Knowledge-Content of Computer Science in Higher Education and the Labor Market.* Washington, DC: U.S. Department of Education and the National Institute for Science Education.

Adelman, C. 1996. "The Thin Yellow Line: Editing and Imputation in a World of Third-Party Artifacts," in Alvey, W. and Jamerson, B. (eds.) *Data Editing Exposition.* Washington, DC: Office of Management and Budget, pp. 173-181.

Agrawal, A. and Bingham, T. 1995. "SLCC [St. Louis Community College]/Southwest Bell Telephone Accelerated Training Program in Telecommunications Technology." In L. P. Grayson (ed.), *Proceedings: 1995 College Industry Education Conference.* Washington, DC: American Society for Engineering Education, pp. 61-63.

Alsalam, N. and Rogers, G. 1991. *The Condition of Education, 1991.* Washington, DC: National Center for Education Statistics. Vol. 2: Postsecondary Education.

American Society for Engineering Education. 1986. *Quality of Engineering Education.* Washington, DC: Author.

Astin, A. W. 1977. *Four Critical Years.* San Francisco: Jossey-Bass.

Astin, A. W. 1993. *What Matters in College?: 'Four Critical Years' Revisited.* San Francisco: Jossey-Bass.

Astin, A. W. and H.S. Astin, 1993. *Undergraduate Science Education: the Impact of Different College Environments on the Educational Pipeline in the Sciences.* Los Angeles: Higher Education Research Institute, U.C.L.A.

Astin, A. W., Hemond, M. K. and Richardson, G. T. 1982. *The American Freshman: National Norms for Fall 1982.* Los Angeles: Higher Education Research Institute, UCLA.

Astin, A. W., Korn, W. S., and Berz, E. R. 1990. *The American Freshman: National Norms for Fall 1990.* Los Angeles: Higher Education Research Institute, UCLA.

Astin, A.W., Tsui, L., and Avalos, J. 1996. *Degree Attainment Rates at American Colleges and Universities: Effects of Race, Gender, and Institutional Type.* Los Angeles: Higher Education Research Institute, UCLA.

Atman, C. J. and Nair, I. 1996. "Engineering in Context: an Empirical Study of Freshmen Students' Conceptual Frameworks." *Journal of Engineering Education*, vol. 85, no. 4, pp. 317-326.

Bailyn, L. 1987. "Experiencing Technical Work: a Comparison of Male and Female Engineers." *Human Relations*, vol. 40, no. 5, pp. 299-312.

Baker, M. 1988. "Retention of New-Era Undergraduate Engineering Students: the Need for Administrative Planning." *Proceedings: 1988 ASEE Annual Conference.* Washington, DC: American Society for Engineering Education, pp. 831-836.

Becher, T. 1989. *Academic Tribes and Territories.* Milton Keynes (UK): Open University Press.

Board of Engineering Education. 1994. *Major Issues in Engineering Education: a Working Paper.* Washington, DC: National Research Council.

Braddock, D. J. 1992. "Scientific and Technical Employment, 1990-2005." *Monthly Labor Review*, vol. 115, no. 2, pp. 28-41.

Brush, S. G. 1991. "Women in Science and Engineering." *American Scientist*, vol. 79. pp. 404-416.

Bucciarelli, L. L. and Kuhn, S. 1997. "Engineering Education and Practice," in S. R. Barley and J. E. Orr (eds.), *Between Craft and Science: Technical Work in U.S. Settings.* Ithaca, NY: Cornell University Press, pp. 210-229.

Burton, L. and Celebuski, C. A. 1995. *Technical Education in 2-Year Colleges.* Arlington, VA: National Science Foundation.

Burton, L. and Celebuski, C. A. 1994. *Undergraduate Education in Electrical, Mechanical and Civil Engineering.* Arlington, VA: National Science Foundation.

Cahalan, M., Farris, E, and White, P. 1990. *Science, Mathematics, Engineering, and Technology in Two-Year and Community Colleges.* Washington, DC: National Science Foundation.

California Postsecondary Education Commission. 1981. *Engineering Education and Licensing in California.* Sacramento, CA.: Author.

California Postsecondary Education Commission. 1986. *Retention of Students in Engineering--a Report to the Legislature in Response to Senate Concurrent Resolution 16 (1985).* Sacramento, CA: Author.

Calvert, M. A. 1976. *The Mechanical Engineer in America, 1830-1910.* Baltimore: Johns Hopkins University Press.

Carr, R. *et al.* 1995. "Mathematical and Scientific Foundations for an Integrative Engineering Curriculum." *Journal of Engineering Education*, vol. 84, no. 2, pp. 137-150.

Chelapati, C.V. (ed.) 1990. *Seismic Design: P.E. (Civil) License Review Manual.* Long Beach, Calif.: Professional Engineering Development Publications, Vol. II.

Coleman, R. J. 1996. "The Engineering Education Coalitions: a Progress Report." *Prism*, vol. 6. no. 1, pp. 24-31.

Constantinople, R.R., Cornelius, R. and Gray, J. 1988. "The Chilly Climate: Fact or Artifact." *Journal of Higher Education*, vol. 59, no. 5, pp. 527-550.

DeVor, R. E. and Kapoor, S.G. 1995. "Agile Manufacturing and Machine-Tool Systems: the NSF/ARPA Machine Tool Agile Manufacturing Research Institute." In L. P. Grayson (ed.), *Proceedings: 1995 College Industry Education Conference.* Washington, DC: American Society for Engineering Education, pp. 53-59.

Dey, E. L., A. W. Astin and W. S. Korn. 1991. *The American Freshman: Twenty-Five Year Trends, 1966-1990.* Los Angeles, CA: Higher Education Research Institute.

Didion, C. J. 1993. "Attracting Graduate and Undergraduate Women as Science Majors: Communicating the Structure and Function of the Scientific Establishment." *Journal of College Science and Teaching*, vol. 22, no. 6, pp. 336-49.

Dorato, P. and Abdallah, C. 1993. "A Survey of Engineering Education Outside the United States: Implications for the Ideal Engineering Program." *Journal of Engineering Education*, vol 82, no. 4, pp. 212-215.

Eisenberg, E. R. 1987. "Characteristics of Highly Rated Associate Degree Programs." *Journal of Engineering Education*, vol. 77, nos. 7 & 8, pp. 735-740.

Ellis, R. A. 1987. "Engineering Enrollments, Fall 1986." *Engineering Education*, vol. 78, no. 3, pp. 155-178.

Ellis, R. A. and Eng, P. 1991. "Women and Men in Engineering." *Engineering Manpower Bulletin*, no. 107. Washington, DC: American Association of Engineering Societies.

Engineering Manpower Commission, 1987. *Engineering and Technology Enrollments: Fall, 1987.* Washington, DC: American Association of Engineering Societies, Inc.

Engineering Manpower Commission, 1991. *Engineering and Technology Enrollments, Fall, 1991.* Washington, DC: American Association of Engineering Societies, Inc.

England, P. 1984. "Explanation of Job Segregation and the Sex Gap in Pay." In *Comparable Worth: Issues for the 80s.* Washington, DC: U.S. Commission on Civil Rights.

Epstein, C. F. 1991. "Constraints on Excellence: Structural and Cultural Barriers to the Recognition and Demonstration of Achievement." In Zuckerman, H., J.R. Cole, and J. T. Bruer, *The Outer Circle: Women in the Scientific Community.* New York: W. W. Norton & Co., pp. 239-258.

Etzkowitz, H. *et al.* 1994. "The Paradox of Critical Mass for Women in Science." *Science*, vol. 21, pp. 45-71.

Felder, R. M. 1995. "A Longitudinal Study of Engineering Student Performance and Retention, IV: Instructional Methods. *Journal of Engineering Education*, vol. 84, no. 4, pp. 361-368.

Felder, R. M., Mohr, P.H., Dietz, E. J. and Baker-Ward, L. 1994. "A Longitudinal Study of Engineering Student Performance and Retention. II. Rural/Urban Student Differences." *Journal of Engineering Education*, vol. 83, no.3, pp. 209-218.

Felder, R. M. *et al.* 1995. "A Longitudinal Study of Engineering Student Performance and Retention. III. Gender Differences in Student Performance and Attitudes. *Journal of Engineering Education*, vol. 84, no.2, pp. 151-164.

Felder, R. M. *et al.* 1993. "A Longitudinal Study of Engineering Student Performance and Retention. I. Success and Failure in the Introductory Course." *Journal of Engineering Education*, vol. 82, no. 1, pp. 15-21.

Fitzpatrick, J. L and Silverman, T. 1989. "Women's Selection of Careers in Engineering: Do Traditional-Nontraditional Differences Still Exist?" *Journal of Vocational Behavior*, vol. 34, pp. 266-278.

Flynn, P.M., Leeth, J.D., and Levy, E.S. 1996. "The Evolving Gender Mix in Accounting: Implications for the Future of the Profession." *Selections*, vol. 12, no. 2, pp. 28-39.

Fox, M. F. 1984. "Women and Higher Education: Sex Differentials in the Status of Students and Scholars." In J. Freeman (ed.) *Women: a Feminist Perspective*. Palo Alto, CA: Mayfield Publishing Co., pp. 238-255.

Ginorio, A.B., Brown, M.D., Henderson, R.S. and N. Cook. 1993. *Patterns of Persistence and Attrition Among Science and Engineering Majors at the University of Washington, 1985-1991*. New York: Alfred P. Sloan Foundation.

Gleick, J. 1987. *Chaos: Making a New Science*. New York: Penguin Books.

Grandy, J. 1998. *Science and Engineering Graduates: Career Advancement and Career Change*. Princeton, NJ: Educational Testing Service. GRE No. 91-14.

Grandy, J. 1995. *Persistence in Science of High-Ability Minority Students, Phase V: Comprehensive Data Analysis*. RR-95-31. Princeton, NJ: Educational Testing Service.

Grandy, J. 1994. *Gender and Ethnic Differences Among Science and Engineering Majors: Experiences, Achievements, and Expectations*. RR-94-30. Princeton, NJ: Educational Testing Service.

Grandy, J. 1989. *Trends in SAT Scores and Other Characteristics of Examinees Planning to Major in Mathematics, Science, or Engineering*. RR-89-24. Princeton, NJ: Educational Testing Service.

Greenfield, L. B., Holloway, E. L. and Remus, L. 1982. "Women Students in Engineering: Are They So Different from Men?" *Journal of College Student Personnel*, vol. 23, pp. 508-514.

Hacker, S. L. 1981. "The Culture of Engineering: Woman, Workplace, and Machine." *Women's Studies International Quarterly*, vol. 4, no. 3, pp. 341-352.

Hacker, S. L. 1983. "Mathematization of Engineering: Limits on Women and the Field." in J. Rothschild (ed.) *Machina ex Dea: Feminist Perspectives on Technology*. New York: Pergamon Press, pp. 38-58.

Hackett, G., Betz, N.E., Casas, J.M., and Rocha-Singh, I.A. 1992. "Gender, Ethnicity, and Social Cognitive Factors Predicting the Academic Achievement of Students in Engineering." *Journal of Counseling Psychology*, vol. 39, no. 4, pp.527-38.

Harmon, L. W. 1989. "Longitudinal Changes in Women's Career Aspirations: Developmental or Historical?" *Journal of Vocational Behavior*, vol. 35, pp. 46-63.

Harrier, J. A. 1996. "The Evolution of the Engineering Community: Pressures, Opportunities, and Challenges." *Journal of Engineering Education*, vol. 85, no. 1, pp. 5-9.

Heckel, R. W. 1996. "Engineering Freshman Enrollments: Critical and Non-Critical Factors." *Journal of Engineering Education*, vol. 85, no.1, pp. 15-22.

Heckel, R. W. 1995. "Disciplinary Patterns in Degrees, Faculty and Research Funding." *Journal of Engineering Education*, vol. 84, no. 1, pp. 31-40.

Heggen, R. J. 1988. "Statics and Dynamics in the Engineering Curriculum." *Journal of Engineering Education*, vol. 78, no. 5, pp. 317-318.

Hendley, V. 1997. "Recruiters Hear A 'Me, Too!' from Community College Students." *ASEE Prism*, April, pp. 21-26.

Henes, R, Bland, M.M., Darby, J., and McDonald, K. 1995. "Improving the Academic Environment for Women Engineering Students Through Faculty Workshops." *Journal of Engineering Education*, vol. 84, no. 1, pp. 59-67.

Heyns, B.L. and Hilton, T.L. 1982. "The Cognitive Tests for High School and Beyond: an Assessment." *Sociology of Education*, vol. 55, no. 2, pp. 89-102.

Hindle, B. 1976. "The Underside of the Learned Society in New York, 1754-1854." In A. Oleson and S. C. Brown (eds.), *The Pursuit of Knowledge in the Early American Republic*. Baltimore: Johns Hopkins University Press, pp. 84-116.

Hilton, T. L. and Lee, V. L. 1988. "Student Interest and Persistence in Science: Changes in the Educational Pipeline in the Last Decade." *Journal of Higher Education*, vol. 59, no. 5, pp. 510-526.

Hinerman, C. 1995. "Unique Needs Drive Innovative Industry-Education Partnerships for Workforce Development and Retraining. In L. P. Grayson (ed.), *Proceedings: 1995 College Industry Education Conference*. Washington, DC: American Society for Engineering Education, pp. 123-126.

Hoffer, T. B. and W. Moore. 1996. *High School Seniors' Instructional Experiences in Science and Mathematics*. Washington, DC: National Center for Education Statistics.

Holland, J. L. 1985. *Making Vocational Choices*. Englewood Cliffs, N.J.: Prentice-Hall.

Howard Hughes Medical Institute. 1997. *Assessing Science Pathways: Tracking Science Education and Careers from Precollege Through Professional Levels*. Chevy Chase, MD: Author.

Humphreys, S. M. and Freeland, R. 1992. *Retention in Engineering: a Study of Freshman Cohorts*. Berkeley, CA: University of California.

Ivey, E. S. 1988. "Recruiting More Women Into Engineering and Science." *Engineering Education*, vol. 78, no. 8, pp. 762-765.

Jagacinski, C. M. and LeBold, W. K. 1981. "A Comparison of Men and Women Undergraduate and Professional Engineers." *Journal of Engineering Education*, vol. 72, no. 3, pp. 213-220.

Kanter, R. M. 1977. *Men and Women of the Corporation*. New York: Basic Books.

Kaufman, D. 1980. "A Survey of Freshman Engineering Curricula." *Journal of Engineering Education*, vol. 70, no. 5, pp. 432-433.

Kroc, R., Howard, R., Hull, P. and Woodard, D. 1997. "Graduation Rates: Do Students' Academic Program Choices Make a Difference." Paper presented to the 1997 Forum of the Association for Institutional Research.

Kunda, G. 1992. *Engineering Culture: Control and Commitment in a High-Tech Corporation*. Philadelphia: Temple University Press.

LeBold, W. K. and Ward, S. K. 1988. "Engineering Retention: National and Institutional Perspectives." *Proceedings: 1988 ASEE Annual Conference*. Washington, DC: American Society for Engineering Education, pp. 843-851.

Linn, M.C. and Hyde, J.S. 1989. "Gender, Mathematics, and Science." *Educational Researcher*, vol. 18, no. 8, pp. 17-19, 22-27.

Lips, H. M. 1993. "Bifurcation of a Common Path: Gender Splitting on the Road to Engineering and Physical Science Careers." *Initiatives*, vol 55. no. 3 (1993), pp. 13-22.

Lips, H. M. and Temple, L. 1990. "Major in Computer Science: Causal Models for Women and Men." *Research in Higher Education*, vol. 31, no. 1, pp. 99-113.

Lodahl, J. and Gordon, G., 1973. "Differences Between Physical And Social Sciences in University Graduate Departments." *Research in Higher Education*, vol. 1, no. 2, pp. 191-213.

Madigan, T. 1997. *Science Proficiency and Course Taking in High School*. Washington, DC: National Center for Education Statistics.

McCormick, A. 1997. *Transfer Behavior Among Beginning Postsecondary Students: 1989-94*. Washington, DC: National Center for Education Statistics.

McIlwee, J. S. and Robinson, J. G. 1992. *Women in Engineering: Gender, Power and Workplace Culture*. Albany, NY: State University of New York Press.

Moller-Wong, C. and Eide, A. 1997. "An Engineering Student Retention Study." *Journal of Engineering Education*, vol. 86, no. 1, pp. 7-16.

National Research Council. 1986. *Engineering Undergraduate Education*. Washington, DC: National Academy Press.

National Research Council. 1994. *Women Scientists and Engineers Employed in Industry*. Washington, DC: National Academy Press.

National Science Foundation. 1986. *Women and Minorities in Science and Engineering*. Washington, DC: Author.

National Science Foundation. 1988. *Women and Minorities in Science and Engineering*. Washington, DC: Author.

Neiner, A. G. and Owens, W. A. 1985. "Using Biodata to Predict Job Choice Among College Graduates." *Journal of Applied Psychology*, vol. 70, pp. 127-136.

Noble, D. 1977. *America by Design: Science, Technology and the Rise of Corporate Capitalism*. New York: Alfred Knopf.

Organization for Economic Cooperation and Development. 1997. *Education at a Glance: OECD Indicators, 1997*. Paris: Author.

Panitz, B. 1996. "Evolving Paths." *Prism* (October), pp. 23-28.

Pascarella, E.T. and P. T. Terenzini, 1991. *How College Affects Students*. San Francisco: Jossey-Bass.

Pavelich, M. J. and Moore, W. S. 1996. "Measuring the Effect of Experiential Education Using the Perry Model." *Journal of Engineering Education*, vol. 85, no. 4, pp. 287-292.

Perrucci, C. 1970. "Minority Status and the Pursuit of Professional Careers: Women in Science and Engineering." *Social Forces*, vol. 49, no. 2, pp. 245-259.

Peters, M., Chisholm, P., and Laeng, B. 1995. "Spatial Ability, Student Gender, and Academic Performance." *Journal of Engineering Education*, vol. 84, no. 1, pp. 69-74.

Pfeffer, J., Leong, A. and Strehl, K. 1977. "Paradigm Development and Particularism: Journal Publication in Three Scientific Disciplines." *Social Forces*, vol. 55, pp. 938-951.

Polachek, S.W. 1978. "Sex Differences in College Major," *Industrial and Labor Relations Review*, vol. 31, pp. 498-508.

Ransom, M. R. 1990. "Gender Segregation by Field in Higher Education." *Research in Higher Education.* Vol., 31, no. 5, pp. 477-491.

Robinson, J. G. and McIlwee, J. S. 1991. "Men, Women and the Culture of Engineering." *The Sociological Quarterly*, vol. 32, no 3, pp. 403-421.

Saigal, A. 1987. "Women Engineers: an Insight Into Their Problems." *Journal of Engineering Education*, vol. 78, no. 3, pp. 194-195.

Sax, L. J. 1994. "Retaining Tomorrow's Scientists: Exploring the Factors that Keep Male and Female College Students Interested in Science Careers." *Journal of Women and Minorities in Science and Engineering,* vol. 1, pp. 45-61.

Schonberger, A. K. 1990. "College Women's Persistence in Engineering and Physical Science." In Keith, S. and P. Keith (eds.), *Proceedings of the National Conference on Women in Mathematics and the Sciences.* St. Cloud, MN: St. Cloud State University, pp. 101-104.

Seely, B. 1993. "Research, Engineering, and Science in American Engineering Colleges: 1900-1960." *Technology and Culture*, vol. 34, pp. 344-386.

Serex, C. P. 1997. "Perceptions of Classroom Climate by Students in Non-Traditional Majors for Their Gender." Paper presented at the Annual Meeting of the Association for the Study of Higher Education.

Seymour, E. and Hewitt, N.M. 1997. *Talking About Leaving: Why Undergraduates Leave the Sciences.* Boulder, Colo.: Westview Press.

Shen, B.S P. 1975. "Science Literacy and the Public Understanding of Science." In S.B. Day (ed.), *Communication of Scientific Information.* Basel (Switzerland): S. Karger, pp. 44-52.

Smart, J. C. 1989. "Life History Influence on Holland Vocational Type Development." *Journal of Vocational Behavior*, vol. 34, pp. 69-87.

Smith, T.M. *et al.* 1996. *The Condition of Education, 1996*. Washington, DC: National Center for Education Statistics.

Sonnert, G. 1996. "Gender Equity in Science: Still an Elusive Goal." *Issues in Science and Technology*, vol. 12, no. 2, pp. 53-58.

Strenta, C, *et al.* 1993. *Choosing and Leaving Science in Highly Selective Institutions: General Factors and the Question of Gender*. New York: Alfred P. Sloan Foundation.

Tobias, S. 1990. *They're Not Dumb, They're Different: Stalking the Second Tier*. Tucson, AZ: Research Corporation.

Tobias, S., D.E. Chubin, and K. Aylesworth. 1995. *Rethinking Science as a Career: Perceptions and Realities in the Physical Sciences*. Tucson, AZ: Research Corporation.

Traunter, J. J., Chou, K. C., Yates, J. K., and Stalnaker, J. "Women Faculty in Engineering: Changing the Academic Climate." *Journal of Engineering Education*, vol. 85, no.1, pp. 45-52.

Tsapogas, J. 1996. *Characteristics of Recent Science and Engineering Graduates: 1993*. Arlington, VA: National Science Foundation.

Vetter, B M. 1988. "Demographics of the Engineering Student Pipeline." *Engineering Education*, vol. 78, no. 8, pp. 735-740.

Wagenaar, T. C. 1984. *Occupational Aspirations and Intended Field of Study in College*. Washington, DC: National Center for Education Statistics.

Warren, J. 1989. "A Model for Assessing Undergraduate Learning in Mechanical Engineering." In C. Adelman (ed.), *Signs and Traces: Model Indicators of College Student Learning in the Disciplines*. Washington, DC: U.S. Department of Education, pp. 65-91.

Wenk, E. 1988. "Portents for Reform in Engineering Curricula." *Journal of Engineering Education*, vol.78, no.11, pp. 99-102.

West, J., L. Diodato, and N. Sandberg. 1984. *A Trend Study of High School Offerings and Enrollments: 1972-73 and 1981-82*. Washington, DC: National Center for Education Statistics.

Whalley, P. 1986. *The Social Production of Technical Work*. Albany, NY: SUNY Press.

Whalley, P. and Barley, S. R. 1997. "Technical Work in the Division of Labor: Stalking the Wily Anomaly." In S. R. Barley and J. E. Orr, *Between Craft and Science: Technical Work in U.S. Settings*. Ithaca: Cornell University Press, pp. 23-52.

White, M. J., Kruczek, T. A., Brown, M. T. and White, G. B., 1989. "Occupational Sex Stereotypes Among College Students." *Journal of Vocational Behavior*, vol. 34, pp. 289-298.

Wolf, L. J. 1987. "The Emerging Identity of Engineering Technology." *Journal of Engineering Education*, vol. 77, nos. 7 & 8, pp. 725-729.

Zuckerman, H., J. R. Cole and J. T. Bruer. 1991. *The Outer Circle: Women in the Scientific Community*. New York: W. W. Norton & Co.

Zussman, R. 1985. *Mechanics of the Middle Class: Work and Politics Among American Engineers*. Berkeley, CA: Univ. of California Press.

Appendix: Technical Notes and Guidance

There are many tables in this document, both in the text and in the notes. Some are derived or constructed from other published sources. Where those sources are a complete census, for example, a survey of engineering enrollments conducted by the American Society for Engineering Education, we don't worry much about samples and weights. Where those sources rely on samples, we must assume that the statistical standards of an agency such as the National Science Foundation, a research organization such as the Higher Education Research Institute at U.C.L.A., or a state agency such as the California Postsecondary Education Commission were observed in the production of data.

But most of the tables in this publication were prepared using special analyses files created from the High School and Beyond/Sophomore Cohort (HS&B/So) longitudinal study of the National Center for Education Statistics, and it is helpful to know something about the statistical standards that lie behind these tables and the decision rules that were used in presenting the data.

The populations in all NCES age-cohort longitudinal studies are national probability samples first drawn when the students were in high school or middle school. In the case of the HS&B/So, the design involved first, a stratified sample of secondary schools with an over-sampling of schools in minority areas, and a random sampling of 10th grade students within those schools. The original sample was then weighted to match the national census of all 10th-graders in 1980 (about 3.7 million people). Each participant carries a weight in inverse proportion to the probability that he or she would be selected by chance. The HS&B/So base year sample was what statisticians call "robust": 28,000. After the base year, every subsequent survey was a subset of the original, and the weights carried by participants are modified accordingly. In the penultimate survey of the HS&B/So in 1992, there were 12,640 respondents out of 14,825 surveyed. The postsecondary transcript file for the HS&B/So has 8,395 cases. These are still very robust numbers. They represent populations in the millions. By the conclusion of any of these longitudinal studies, a student is carrying a half-dozen different weights, depending on what question is asked.

For the High School and Beyond cohort, for example, I used four different weights in the tables in this study: a "senior year" weight for a question such as the relationship between the highest level of mathematics studied in high school and whether someone eventually earns a bachelor's degree; a "primary postsecondary transcript weight" for analyses of degree attainment and major fields along the engineering path; a "secondary transcript weight" for any question that would be compromised if students with incomplete records were included; and a "final weight" for 1991 labor market experience analyses.

More important are issues of standard errors of measurement and significance testing. What you see in the tables are estimates derived from samples. Two kinds of errors occur when samples are at issue: errors in sampling itself, particularly when relatively small subpopulations are involved, and non-sampling errors. Non-sampling errors are serious matters. Good examples would include non-response to specific questions in a survey or missing college transcripts. Weighting will not address the panoply of sources of non-sampling errors.

The effects of sampling and non-sampling errors ripple through databases, and, to judge the accuracy of any analysis, one needs to know those effects. When the unit of analysis is the student, this is a straightforward issue. When we ask questions about combinations of institutions attended (table 7), bachelor's degree completion rates by transfer status (table 9), highest level of mathematics studied in high school (table 15), or 1991 earnings (table 23), we are asking questions about non-repetitive behaviors of people who were sampled. To judge comparisons in these cases we use the classic "Student's *t*" statistic that requires standard errors of the mean. But because the longitudinal studies were not based on simple random samples of students, the technique for generating standard errors involves a more complex approach known as the Taylor series method. For the descriptive statistics in this report, a proprietary program incorporating the Taylor series method, called STRATTAB, was used.

It is important to note that STRATTAB will provide neither estimates nor standard errors for any cell in a table in which the unweighted N is less than 30. For those cells, the program shows "LOW N." Table 6 on page 21 illustrates the frequency of LOW N cells that occur when one is making multiple comparisons among categories of an independent variable.

Most of the tables in this monograph include standard errors of the estimates and/or an indication of which comparisons in the table are significant at the $p \leq .05$ level using the classic "Student's *t*" test. The text often discusses these cases, and, when appropriate to the argument, offers the *t* statistic. A reader interested in comparing categories of a dependent variable that are not discussed can use the standard errors and employ the basic formula for computing the "Student's *t*":

$$t = (P_1 - P_2) / \quad (se_1^2 + se_2^2)$$

where P_1 and P_2 are the estimates to be compared and se_1 and se_2 are the corresponding standard errors. If, in this case, $t \geq 1.96$, you have a statistically significant difference such that the probability that this observation would occur by chance is less than 1:20. In the case of multiple comparisons, the critical value for *t* rises following the formula for Bonferroni Tests: if *H* comparisons are possible, the critical value for a two-sided test is $Z_{(1-.05/2H)}$.

There are some tables for which neither standard errors nor *t* statistics are offered. These are rare. They involve cases where an HS&B/So distribution is compared to a distribution from secondary source (e.g. tables 11 and 15), and complex tables with many cells (e.g. tables 10 and 12, that are actually based on calculations involving over 2,000 cells) and where there is simply too much already on the page. Even if standard errors could be computed for tables 10 and 12. they would only distract the reader's attention from the point of the tables.

ISBN 0-16-049551-2